AA001074

MATERIALS RESEARCH SOCIETY
SYMPOSIUM PROCEEDINGS VOLUME 681

Wafer Bonding and Thinning Techniques for Materials Integration

April 16 – 20, 2001
San Francisco, California, USA

Printed from e-media with permission by:

Curran Associates, Inc.
57 Morehouse Lane
Red Hook, NY 12571
www.proceedings.com

ISBN: 1-55899-617-6

Some format issues inherent in the e-media version may also appear in this print version.

CAMBRIDGE UNIVERSITY PRESS
Cambridge, New York, Melbourne, Madrid, Cape Town,
Singapore, São Paulo, Delhi, Tokyo, Mexico City

Cambridge University Press
32 Avenue of the Americas, New York, NY 10013-2473, USA

www.cambridge.org

Materials Research Society
506 Keystone Drive, Warrendale, PA 15086
http://www.mrs.org

©Materials Research Society 2001

This publication is in copyright. Subject to statutory exception
and to the provisions of relevant collective licensing agreements,
no reproduction of any part may take place without the written
permission of Cambridge University Press.

First published 2001

CODEN: MRSPDH

ISBN: 1-55899-617-6

Cambridge University Press has no responsibility for the persistence or
accuracy of URLs for external or third-part Internet Web sites referred to
in this publication and does not guarantee that any content on such Web sites
is, or will remain, accurate or appropriate.

Additional copies of this publication are available from:

Curran Associates, Inc.
57 Morehouse Lane
Red Hook, NY 12571 USA
Phone: 845-758-0400
Fax: 845-758-2634
Email: curran@proceedings.com
Web: www.proceedings.com

TABLE OF CONTENTS

1 **Integration of Materials and Device Research Enabled by Wafer Bonding and Layer Transfer**
Q.-Y. Tong

24 **Plasma Induced Chemical Changes at Silica Surfaces During Pre-Bonding Treatments**
Darren M. Hansen, C.E. Albaugh, Peter D. Moran, and T. F. Kuech

30 **Molecular dynamics simulations of wafer bonding**
Kurt Scheerschmidt

39 **Wafer Bonding of Silicon Carbide and Gallium Nitride**
Jaeseob Lee, T. E. Cook, E. N. Bryan, J. D. Hartman, R. F. Davis, and R. J. Nemanich

45 **TEM measurement of hydrogen pressure within a platelet**
J. Grisolia, G. Ben Assayag, B. de Mauduit, A. Claverie

51 **Blistering on Silicon Surface Caused by Gettering of Hydrogen on Post-Implantation Defects**
A.Y. Usenko, and W.N. Carr

57 **Electrical characterisation of UHV-bonded silicon interfaces**
A. Reznicek, S. Senz, O. Breitenstein, R. Scholz, and U. Gösele

63 **Ge Layer Transfer To Si For Photovoltaic Applications**
James M. Zahler, Chang-Geun Ahn, Shahrooz Zaghi, and Harry A. Atwater

69 **Anodic Bonding at Room Temperature**
Volker Baier, Andreas Gebhardt, and Stefan Barth

73 **Si/GaAs heterostructures fabricated by direct wafer bonding**
Viorel Dragoi, Marin Alexe, Manfred Reiche, Ionut Radu, Erich Thallner, Christian Schaefer, and Paul Lindner

79 **SiC-Si Grooved Surface Bonding**
Tatiana S. Agrunova, Igor V. Grekhov, Lioudmila S. Kostina, Alexander G. Tur'yanskii, Igor V. Pirshin, Ilya R. Prudnikov, and Konstantin B. Kostin

85 **Direct Bonding of Silicon Wafers with Simultaneous Dopant Diffusion**
Igor V. Grekhov, Tatiana S. Agrunova, Lioudmila S. Kostina, Natalia M. Shmidt, Helmut Föll, and Konstantin B. Kostin

91 **Thermomechanical stress in silicon on quartz wafer bonding and Smart Cut® process**
Yu-Lin Chao, Qin-Yi Tong, Ulrich M. Gösele1

96 **Anodic Bond Quality and impact on Pressure Sensor Long-Term Stability**
Henry Allen, Kamrul Ramzan, Jim Knutti, Carl Ross, Tim Milliman, Jeff Frye

101 **A Novel Method of Fabricating SiC-On-Insulator Substrates for Use in MEMS**
Hung-I Kuo, Christian Zorman, and Mehran Mehregany

107 **Wafer Bonding between Magnetic Garnet and Lithium Niobate for Semi-Leaky Isolator**
Hideki Yokoi, Tetsuya Mizumoto, and Masafumi Shimizu

112 **Integration of InGaN-based Optoelectronics with Dissimilar Substrates by Wafer Bonding and Laser Lift-off**
William S. Wong, Michael Kneissl, David W. Treat, Mark Teepe, Naoko Miyashita, and Noble M. Johnson

121 **Ultrathin Slices of Ferroelectric Domain-Patterned Lithium Niobate by Crystal Ion Slicing**
David A. Scrymgeour, Venkat Gopalan, Tony E. Haynes, and Miguel Levy

127 **Fabrication of Three-Dimensional Photonic Crystal by Wafer Fusion Approach**
Noritsugu Yamamoto, Katsuhiro Tomoda, and Susumu Noda1

134 **Multiple Wafer Bonding for MEMS Applications**
M. Reiche , M. Haueis , J. Dual , C. Cavalloni , and R. Buser

140 **Packaging Of Ultrathin Semiconductor Devices Through The ELO Packaging Process**
Mike Sickmiller

146 **A Novel Ultra-miniature catheter tip pressure sensor fabricated using silicon and glass thinning techniques**
Henry Allen, Kamrul Ramzan, Jim Knutti, and Stan Withers

152 **Bonding, Splitting and Thinning by Porous Si in ELTRAN® ; SOI-Epi Wafer™**
Kenji Yamagata and Takao Yonehara

162 **Atomic-Layer Cleaving and Non-contact Thinning and Thickening for Fabrication of Laminated Electronic and Photonic Materials**
Michael I. Current, Shari N. Farrens, Martin Fuerfanger, Sien Kang, Harry R. Kirk, Igor J. Malik, Lucia Feng, Francois J. Henley,

171 **Orientation and Boron Concentration Dependence of Si Layer Transfer by Mechanical Exfoliation**
Kimmo Henttinen, Tommi Suni, Arto Nurmela, Veli-Matti Airaksinen, Ilkka Suni, and S.S Lau

177 **Structured monocrystalline Si thin-film modules from layer-transfer using the porous Si (PSI) process**
Auer Richard and Brendel Rolf

183 **Transfer and handling of thin semiconductor materials by a combination of wafer bonding and controlled crack propagation**
J. Bagdahn1, D. Katzer , M. Petzold, M. Wiemer, M. Alexe, V. Dragoi, U. Goesele

189 **Gettering Control at Bonding Interface in ELTRAN®**
Kazutaka Momoi, Masataka Ito, Nobuhiko Sato, Noriaki Honma and Takao Yonehara

195 **Characterization of Optical Lifetime in Silicon-on-Insulator Wafers by Photoluminescence Decay Method**
Shigeo Ibuka, Michio Tajima, and Atsushi Ogura

Mat. Res. Soc. Symp. Proc. Vol. 681E © 2001 Materials Research Society

INTEGRATION OF MATERIALS AND DEVICE RESEARCH ENABLED BY WAFER BONDING AND LAYER TRANSFER

Q.-Y. Tong

Wafer Bonding Lab., Research Triangle Institute and Ziptronix, RTP, NC 27709

Wafer bonding and layer transfer technology has emerged as a versatile approach for integrated research and development of materials and devices. It does not only provide a flexible way to prepare integrated materials but also breaks the barrier between materials science and device engineering since it appears to be one of the fundamental technologies for 3-D device fabrication and for integration of partially or fully processed dissimilar functional layers. The device performance can be significantly enhanced when both sides of device layers can be processed such as in the double gate CMOS and in HBTs. The processed device layers can be considered as unique materials layers and can also be integrated. The integration of dissimilar integrated circuits provides a promising solution to realize micro-systems or system on a 3-D chip in which multi-functional layers are interconnected (3-D SOC). The main issues are to develop methods that form strong bond between required materials at low temperatures and to cut the device layer without compromising the device integrity by using VLSI compatible processes. The advances in low temperature bonding of hydrophilic and hydrophobic silicon, compound semiconductors and other materials in ambient are reviewed, and application examples are discussed.

Introduction

Wafer direct bonding is a technology that allows wafers to be bonded at room temperature without using any adhesive or external forces, therefore is not prone to introduce stress and inhomogeneity. Wafer direct bonding is also different from the popular anodic bonding that employs heating and electric field and requires at least one of the bonding wafers to be a glass wafer or a wafer covered by a glass layer containing mobile ions. Wafer direct bonding technology is based on understanding the question of why broken pieces of any solid material usually can not be reversibly re-joined at room temperature in ambient even if the mating surfaces are perfectly complementary. It was understood that the main factors that prevent reversible rejoining appear to be the changes of the surfaces immediately after separation including surface reconstruction, roughening, adsorption, oxidation and contamination that reduce the surface energy significantly or prevent the surfaces from coming in close proximity. Based on this understanding, wafer bonding technology was introduced such that two wafers with surfaces that are sufficiently smooth, flat and clean can bond to each other without any adhesive or external forces at room temperature in ambient air [1, 2]. However, it was found that the standard wafer direct bonding was attributed to relatively weak van der Waals forces and subsequent annealing at high temperature is required to achieve a strong bond.

To monolithically combine a variety of materials to form integrated materials for integrated circuits (ICs) with enhanced performance has been one of the driving forces to develop wafer bonding technology as the scaling limits of the conventional bulk silicon device are approaching. For instance, compared with bulk silicon, CMOS ICs using silicon-on-insulator (SOI) substrates have shown 30% speed gain at the same voltage and are 3-10 times faster at same power. In order to fully explore the potential provided by wafer bonding technology, a high degree of customization of integrated materials is inevitable. Since the device structure and performance are closely related to the materials design, new devices may require novel materials, and new materials opportunities can create unique devices. The joint efforts of materials developers and device engineers may be essential to the further development of the semiconductor industry.

Moreover, it has been recognized recently that device performance can be significantly enhanced by using wafer bonding and layer transfer technology. For instance, in the double gate CMOS [3, 4] source/drain resistance is reduced and the driving current can be increased by ~ 40% or the speed can be enhanced by ~100%. Also wafer bonding and layer transfer allow fabrication of unique HBT integrated circuits (Hetro-junction Bipolar Transistor) [5] in which both sides of the HBT device layers are processed. The HBT extrinsic parasitics are significantly reduced and f_T in excess of 100 GHz with high current gain of ~80 and very low leakage have been achieved. For system-on-a chip (SOC) preparation, a variety of functions are required on a chip. Many functions are usually best made from their respective materials other than silicon. Therefore, integrated materials that combine thin films of dissimilar materials in a single silicon wafer are highly desirable. In order to bond dissimilar materials having different thermal expansion coefficients at wafer level, low temperature bonding is essential. Low temperature bonding is also crucial for materials having low decomposition temperatures or being temperature sensitive even though they are thermally matched. When one wafer of a bonded pair is thinned to a thickness less than the respective critical value for the specific materials combination, the generation of misfit dislocations in the layer and sliding or cracking of the bonded pairs during subsequent thermal processing steps can be avoided.

The design of the mix of processes needed to produce different functions on the same chip can be difficult and hard to optimize. Also, the resultant SOC chips may get too large leading to a low yield. Therefore, a promising alternative approach appears to be to interconnect different ICs that are fully processed and tested to form stacking ICs [6, 7] by wafer adhesive bonding and layer transfer. Since wafer direct bonding and layer transfer is a VLSI (Very Large Scale Integration) compatible, highly flexible and manufacturable technology, using that to form 3-D SOC is highly preferable. The 3-D SOC approach can be seen as the integration of integrated circuits to form a system on a chip, and is a natural evolution of current 2-D integrated circuit technology that interconnects transistors to form circuit on a chip (COC). The 3-D SOC approach is also complementary to the materials integration method because the processed functional layers can be considered as unique dissimilar materials layers.

In this paper, the generic nature of wafer bonding and layer transfer is described. Recent innovative wafer bonding technologies are introduced, especially that of room temperature covalent bonding and low temperature epitaxial bonding. Some samples of integrated materials and devices will be presented.

Bonding anything to anything

--Generic nature of wafer bonding

When two wafers are bonded, the energy required to separate two bonding wafers can be expressed as the product of the number of the bonds formed on a unit area (bond density), n and the energy of each bond E_b. Commonly, specific surface energy γ is used to measure the bonding energy:

$$\gamma = \frac{1}{2} n E_b \qquad (1)$$

Since only molecules that are in sufficient proximity can form bonds between each other, the number of bonds is determined by the smoothness of the contacting surfaces. The adequate level of smoothness of the mating surfaces for bonding depends on the surface bonding species that determine the distance over which the inter-molecular forces are effective.

If non-polar molecular groups terminate the mating surfaces, the attraction force is mainly dispersion force resulting from the nonzero average value of the square of the temporary dipole moment due to charge distribution fluctuations. For the dispersion force to be effective, the peak-valley distance of the surface roughness should be in the same order of the lattice constants of most materials, i.e. around 3-5.5 Å [8] corresponding to the root mean square micro-roughness (RMS) of ~0.5 Å.

For the bonding surfaces that are terminated with polar groups, especially those with hydrogen atoms such as H-F, H-O, and H-N, a strong form of dipole-dipole attraction is present, termed hydrogen bonding. For many polar molecules such as HOH HF or NH_3, a cluster of two or three polar molecules is energetically more favorable than isolated molecules [9]. The linkage of the polar molecules may form a bridge between the two mating surfaces leading to a "long-range" intermolecular force. The requirement of surface smoothness for room temperature bonding is thus greatly eased. For surfaces terminated by OH, NH and FH, the root mean square micro-roughness (RMS) can be up to ~5 Å [10].

Even when the mating surfaces are sufficiently smooth, the macroscopic flatness non-uniformity or waviness of the wafer surfaces can result in a gap (unbonded area) generation at the interface when two such surfaces are brought into contact. However, according to the theory of small elastic deflection of a thin plate, if the bonding energy is sufficiently high, gaps with height 2h and extension 2R caused by waviness of the bonding surfaces can be closed by elastic deformation of the two wafers. The closure of the gaps results in a gain in bonding energy because new bonds at the newly closed interface can develop. Assuming bonding of two identical wafers with thickness of t_w, for $R > 2t_w$ the bonding energy γ required to close the gaps is given by [10]:

$$\gamma > \frac{2\,E't_w^3h^2}{3\,R^4} \qquad (2)$$

where E' is given by $E/(1-v^2)$ with E being Young's modulus and v Poisson's ratio.

According to Eq.(1), in order to achieve a successful bonding, the mating surfaces must be sufficiently clean and terminated only with desired bonding species that are maximized in density. In another word, surface contaminants can significantly reduce the surface reactivity as well as the resultant bonding energy.

Moreover, the trapped interface particles can result in bubbles many orders of magnitude larger than the particles. A particle of about 1 μm diameter typically leads to a bubble with a diameter of about 5000 times larger (0.5 cm) at interface of room temperature bonded standard 4-in Si wafers with a thickness of 525 μm.

The trapped contaminants at the bonding interface can generate bubbles during subsequent annealing after bonding. It has been found that hydrocarbons at the mating surfaces act as the nucleation sites for interface bubbles [11] and hydrogen is the main content in the bubbles of Si/Si bonded pairs [12].

As mentioned above, room temperature adhesion is attributed to the cluster of two to three wafer molecules bridging the mating surfaces terminated by polar molecules such as OH groups. These water molecules must be removed by dissolving into the surrounding materials before the hydrogen bonding between the OH groups on the mating surfaces can be formed. It appears that the polymerization reaction between the OH groups starts immediately even at room temperature when they are in sufficient proximity [13]:

$$Si\text{-}OH + HO\text{-}Si \leftrightarrow Si\text{-}O\text{-}Si + HOH \qquad (3)$$

The above reaction is reversible at $T < \sim 425^{o}C$ and most of the covalent Si-O-Si bonds can not remain at the bonding interface. Even at $T > 425^{o}C$, the interface may still have water molecules. In order for the bonding energy to achieve the bulk fracture energy, an annealing, typically at higher than $1000^{o}C$ for Si/Si bonding is therefore necessary to drive out the water molecules and to close the micro-gaps.

If epitaxial or hetero-epitaxial-like interface is desired, the bonding surfaces must be free of oxide. Silicon wafers are typically dipped in dilute HF to remove any oxide layer followed by room temperature bonding. The HF treated silicon surfaces are terminated mostly by hydrogen but also some fluorine and are called hydrophobic surfaces. The cluster of hydrogen bonded two or three HF molecules bridges the mating surface. Temperatures higher than $700^{o}C$ are usually employed to release hydrogen from

Si-H and to remove H_2 molecules so that the covalent bonds at the bonding interface can be formed [14]:

$$Si\text{-}H + H\text{-}Si \rightarrow Si\text{-}Si + H_2 \qquad (4)$$

Since the van der Waals forces are ubiquitous and are applicable to almost all substances that are in intimate contact, two solid-state plates of almost any materials can be bonded to each other at room temperature.

Polymerization of hydrogen bonded hydrophilic surfaces at low temperatures to form strong covalent bonds appears to be also a rather generic effect and can be applied to bonding of many different materials. For metals or semiconductor M with relatively high electronegativity such as those in group III or higher and some transition metals, their oxides show small ionic character and their hydroxides may also be able to polymerize at low temperatures to form a strong covalent bond [13]:

$$M\text{-}OH + HO\text{-}M \rightarrow M\text{-}O\text{-}M + HOH \qquad (5)$$

Wafer bonding provides a high degree of flexibility in materials integration. Differences in wafer size, thickness, crystallographic structure, orientation and quality, doping type and profile, materials combination and structure are not obstacles for wafer bonding.

Transferring layer from anything onto anything

-Generic nature of layer cutting

In most of wafer bonding applications, thinning of one wafer of a bonded pair to a thickness less than the respective critical value for the dissimilar materials combination is essential for avoiding the generation of misfit dislocations in the layer and for preventing sliding, debonding and cracking during subsequent thermal processing steps. Main technologies of uniformly thinning large wafers include etch-back using a built-in etch-stop layer [15], hydrogen implantation induced layer splitting [16] and layer mechanical peeling [17, 18].

The etch-back method is restricted to materials that can have a buried etch stop layer in the wafer, and the wafer substrate and the etch stop layer can be selectively etched away consecutively. The most promising approach for layer transfer is to embrittle a region underneath the wafer at a depth corresponding to the layer thickness. Since hydrogen is known as a bond-breaking element, the embrittlement of the region may most conveniently be achieved by hydrogen ion implantation. The less strongly connected cleavage planes in single crystalline materials may relatively easily be embrittled, especially in brittle materials. The wafer with a buried embrittled region is then bonded to a desired substrate on which the layer is to be transferred. The critical requirement for a successful layer transfer is that the bonding energy at the bonding interface γ_b is higher than the fracture energy in the embrittled region γ_e at the temperature at which layer transfer is to take place:

$$\gamma_b > \gamma_e \qquad\qquad (6)$$

Layer transfer onto a desired substrate can be achieved by layer splitting induced by internal gas pressure in the implanted peak region at elevated temperatures or by forced peeling by using a gas or water jet at room temperature. This approach usually results in a thickness uniformity of transferred layers in the 100 Å range. Moreover, this technology is generic in nature and may be applied to almost all-crystalline materials at a reasonable hydrogen dose.

For an effective embrittlement by hydrogen implantation the defect density and profile in the peak region should be adequate so that sufficient number of platelets can be formed and become the dominating H trapping sites. Due to many competing sites of H trapping too high a defect density such as in an amorphous structure will prohibit platelet formation at reasonable H dose. Too low a defect density will reduce hydrogen solubility and platelet formation probability.

In order to obtain an adequate defect density and profile at as low as possible implant doses, H implantation can be performed with wafer at elevated temperature [19]. When implantation is performed at elevated temperatures, some implant damage can be removed by the dynamic annealing and damage accumulation can be prevented due to the enhanced mobility of point defects leading to an increased recombination or annihilation process. The fewer competing defects for H trapping outside the implanted H peak and an increased H mobility during H implantation result in more platelets at the implanted H peak region. However, the H implantation temperature must be within a temperature window that is specific for each material [19]. This method has been applied to many materials including Si, SiC, GaAs, InP, c-cut sapphire, $LaAlO_3$, GaN, $LiNbO_3$ and $Pb_{0.91}La_{0.09}(Zr_{0.6}Ti_{0.4})O_3$.

Low temperature covalent bonding

One of the key technologies to accomplish integration of layers of dissimilar materials is low temperature bonding. Using pre-heating [20] and/or atomic hydrogen etching [21] or Ar sputtering [22] in UHV (ultra high vacuum) to remove oxide and the adsorbed contaminants from the mating surfaces and to increase the surface energy, low or room temperature covalent bonding of a variety of materials has been demonstrated.

Sputter deposited metals such as Ti, Pt, Au, Pd have also been bonded in-situ at room temperature in UHV [23]. The more flexible and manufacturable wafer direct bonding methods that achieve a strong bond between mating wafers in ambient at low or room temperature are highly desirable.

The polymerization of hydrogen bonded hydrophilic surfaces to form strong covalent bonds at room temperature is feasible [13] as shown in Eq. (3). Since the polymerization reactions are reversible at low temperatures, the key to obtain stable covalent bonds at bonding interface is to remove the by-products at low or room temperature [13].

Oxygen or argon plasma treatment of silicon wafers followed by water rinse and spin-dry prior to bonding results in a bonding energy of ~1000 mJ/m^2 at room temperature after 24 h storage in air [24]. It was speculated that the plasma treatment made the mating surfaces easier to absorb the water molecules generated from the polymerization reaction.

We have developed a room temperature wafer direct bonding technology that is performed in ambient air and can result in bonding energy of ~2500 mJ/m^2 at room temperature [25]. Both dry and wet technology can be used to modify the mating surfaces before bonding. The microscopic image (100X) of a fractured silicon island on oxide covered AlN wafer surface that is forcibly separated from a room temperature bonded silicon/AlN pair is shown in Fig. 1. The room temperature covalent bonding opens up many opportunities in wafer bonding applications including integration of thin layers that are temperature sensitive. Employing this technology, thin silicon device layers of whole wafer size were transferred onto quartz (Si/quartz), BeO (Si/BeO) and AlN (Si/AlN) substrates. Thin InP device films of whole wafer size were also transferred onto Si (InP/Si), AlN (InP/AlN), sapphire (InP/sapphire) and SiC (InP/SiC) substrates. Fig. 2 shows the IR photo of an InP device layer transferred onto a Si device wafer and interconnection between them have been made. In addition to bonding at wafer level, in order to realize the 3-D SOC from existing ICs, we have been able to bond individual IC chips to different chips on a host IC wafer by the room temperature bonding technology. The substrate of the individual IC chips is then removed and interconnects are formed between the bonded individual IC chip and the chip on the host wafer by a VLSI compatible process. A 3" double-side processed 100GHz SIHBT (symmetrical intrinsic hetero-bipolar transistor [5]) InP integrated circuit on AlN substrate has also been developed using a manufacturable room temperature wafer bonding and layer transfer technology. A schematic of the process flow of SIHBT wafer is given in Fig. 3. Because all extrinsic InP materials except those absolutely necessary for device operation are removed, and the emitter and collector can be aligned, the SIHBT integrated circuits are potentially very low power and very high speed integrated circuits. This method also allows the integration of InP HBTs with silicon circuits on the same chip for mixed signal processing.

When the bonding interface is in the device active region rather than only for mechanical support it is usually required to be electric and thermal conducting and/or optical transparent. Epitaxial or hetero-epitaxial-like bonding interface is therefore essential. It is typically achieved by removing native oxide layer from the bare wafers prior to room temperature contacting followed by high temperature annealing. In silicon case, an annealing at higher than 700°C is required. For III V compound wafer bonding to

12

each other or to other materials such as silicon wafers, the annealing at > 600°C in hydrogen with an external pressure of ~ 100 g/cm^2 is commonly employed. Due to the large thermal mismatch between the dissimilar materials, only small pieces, typically 1 cm x 1 cm can be bonded. Although UHV bonding can achieve epitaxial or heteroepitaxial-like bonding interface at low or room temperature bonding as mentioned above, it usually requires pre-heating (~ 600°C) the wafers prior to bonding, or applying pressure during bonding in addition to UHV conditions.

Recently, a low temperature epitaxial or heteroepitaxial-like bonding method has been suggested [26]. From reaction (4) it is clear that the key to achieve high bonding energy of bonded hydrophobic silicon pairs at low temperature is to lower the temperature at which the hydrogen can be released from Si-H$_x$ groups and be removed from the bonding interface.

From a stand-alone silicon wafer surface dipped in HF the release of hydrogen was demonstrated to start at about 367°C from Si-H$_2$ and 447°C from Si-H in UHV [27]. It is energetically favorable for the released hydrogen atoms to form hydrogen molecules. Hydrogen molecules become mobile in silicon at temperatures higher than 500°C [28]. It has been shown that to deplete hydrogen from silicon bonding interface an annealing temperature higher than ~700°C is required.

It is known that boron in silicon and other semiconductors is strongly passivated by hydrogen resulting in B-H complexes. Hydrogen can be released from B-H complexes starting from ~160°C [29]. It has been reported that the neighboring boron appears to weaken the Si-H$_x$ bonds leading to reduce the activation energy of release of hydrogen from the trapping center. It is evidenced that adding a small amount of boron atoms in a hydrogen-implanted region in silicon or other semiconductors can significantly lower the blistering temperature [30].

We have developed low temperature epi-like bonding technology that operates at wafer level in ambient conditions based on the above understanding. A very thin boron doped layer (~ 10 Å) was introduced on the surfaces of mating wafers by common doping techniques. The bulk silicon fracture energy of ~2500 mJ/m^2 of hydrophobic Si/Si pairs was achieved at ~350°C. It can also be applied to hetero-epitaxial-like bonding of dissimilar materials. Bulk InP fracture surface energy of 3" hydrophobic InP/Si wafer pairs was achieved at ~200°C. The comparison of the bonding energy as a function of annealing temperature of bonded hydrophobic InP/Si wafers that were bonded by standard and the epi-like bonding procedures, respectively is shown in Fig. 4. After annealing at 150°C, the InP/Si pairs can withstand lapping, etching, annealing and other process steps. No oxygen was detected at the bonding interface as measured by the SIMS (Secondary Ion Mass Spectroscopy) profiles as shown in Fig. 5. Because of the rough InP surface induced by chemical etching to thin the layer for the SIMS measurement, the bonding interface region is extremely broadened in the SIMS profile.

Conclusion

It appears that two solid-state plates of almost any materials can be directly bonded to each other at room temperature provided that their surfaces are sufficiently smooth, flat and clean. Combined with the generic layer cutting methods, transfer of almost any layer onto any substrate may be possible. Not only integrated materials can be made but also 3-D devices and 3-D SOC may be developed. The major challenges appear to be bonding and layer cutting at low temperature and the joint efforts of materials researchers and device engineers. Innovative low temperature wafer bonding technologies have shown that room temperature covalent bonding and low temperature epitaxial or hetero-epitaxial-like bonding are feasible.

Acknowledgements

The author is grateful for the valuable assistance and the support of the Center of Semiconductor Research at Research Triangle Institute and Ziptronix Inc.

References

1. J. B. Lasky, *Appl. Phys. Lett.,* 48 (1986) 78-80.

2. M. Shimbo, K. Furukawa, K. Fukuda and K. Tanzawa, *J. Appl. Phys.* 60 (1986) 2987-2989.

3. S. Nakamura, H. Horie, K. Asano, T. Fukano and S. Sasaki, *Tech. Dig. Int. Electron. Devices Meet.,* 889 (1995).

4. J.H. Lee, G. Taraschi, A. Wei, T. Langdo, E. Fitzgerald and D. Antoniadis, *Tech. Dig. Int. Electron. Devices Meet.*, 71 (1999).

5. P.M. Enquist and D.B. Slater, Jr., U.S. Patent 5,318,916 (1994).

6. Y. Hayashi, S. Wada,K. Kajiyana, K. Oyama, R. Koh, S takahashi and T. Kunio, *Symp. VLSI Tech. Dig.* 95 (1990)

7. P. Ramm, R. Buchner, U.S. Patent 5,563,084 (1995)

8. K. Stokbro and E. Nielsen, *Phys. Review* B, 58(1998) 1618

9. J. Del Bene and J.A. Pople, *J. Chem. Phys.* 52(1970) 4858.

10. Q.-Y. Tong and U. Gösele, *Semiconductor Wafer Bonding: Science and Technology*, John Wiley & Sons, New York, 1999.

11. K. Mitani, V. Lehmann, R. Stengl, D. Feijoo, U. Gösele and H. Z. Massoud, *Jpn. J. Appl. Phys.* 30 (1991) 615.

12. S. Mack, H. Baumann, U. Gösele, H. Werner, and R. Schlögl, *J. Electrochem. Soc,* 144 (1997) 1106.

13. Q.-Y. Tong and U. Goesele, *J. Electroch. Soc.*, 142 (1995) 3975.

14. Q.-Y. Tong, E. Schmidt, U. Gösele and M. Reiche, *Appl. Phys. Lett.* 64(1994) 625.

15. W. P. Maszara, B.-L. Jiang, A. Yamada, G.A. Rozgoni, H. Baumgart, and A.J.R. De Kock, *J. Appl. Phys.*, 69(1991) 257.

16. M. Bruel, *Electronics Lett.*, 31(1995) 1201.

17. W.G. En, I.J. Malik, M.A. Bryan, S. Farrens, F.J. Henley, N.W. Cheung and C. Chan, *Proc. IEEE Int. SOI Conf.* P8CH36199, p. 163 (1998).

18. K. Sakaguchi, K. Yanagita, H. Kurisu, H. Suzuki, K. Ohmi and T. Yonehara, *Proc. of 1999 IEEE Int. SOI Conf.* 99CH36345, p.111 (1999).

19. Q.-Y. Tong and U. Goesele, *Advanced Materials*, 17 (1999) 1.

20. U. Gösele, H. Stenzel, T. Martini, J. Steinkirchner, D. Conrad and K. Scheerschmidt, *Appl. Phys. Lett.*, , 67 (1995) 3614.

21. T. Akatsu, A. Ploessl, H. Stenzel and U. Goesele, *J. Appl. Phys.* 86 (1999) 7146.

22. H. Takagi, R. Maeda, N. Hosoda and T. Suga, *Appl. Phys. Lett.*, 74 (1999) 2387.

23. T. Shimatsu, R.H. Mollema, D. Monsma, E.G. Keim and J.C. Lodder, *J. Vac. Technol. A* 16(4) (1998) 2125.

24. S. Bengtsson and P. Amirfeiz, *J. Electronic Materials*, 29 (2000) 909.

25. Q.-Y. Tong, et al, US Patent pending.

26. Q.-Y. Tong, US Patent pending.

27. P. Gupta, V. Colvin and S. Geroge, *Phys. Rev. B*, 37(1988) 8234.

28. W.K. Chu, R.H. Kastl, R.F. Lever, S. Mader and B.J. Masters, *Phys. Rev. B*, 16 (1987) 3851.

29. S.J. Pearton, J.W. Corbett and T.S. Shi, *Appl. Phys. A*, 43 (1987) 153.

30. Q.-Y. Tong, R. Scholtz, U. Goesele, and T.-H. Lee, L.-J. Huang, Y.-L. Chao, and T.Y. Tan, *Appl. Phys. Lett.* 72 (1998) 49.

Figure Captions

Figure 1 Microscopic image (100X) of a fractured silicon island on oxide covered AlN wafer surface that is forcibly separated from a room temperature bonded silicon/AlN pair

Figure 2 IR photo of an InP device layer transferred onto a Si device wafer and interconnection between them that have been made

Figure 3 Schematic of process flow of SIHBT wafer fabricated by room temperature wafer bonding and layer transfer

Figure 4 Comparison of the bonding energy as a function of annealing temperature of bonded hydrophobic InP/Si wafers that were bonded by standard and the epi-like bonding procedures, respectively

Figure 5 SIMS (Secondary Ion Mass Spectroscopy) profiles of Si, P and O across the bonding interface of 150C bonded InP/Si wafer

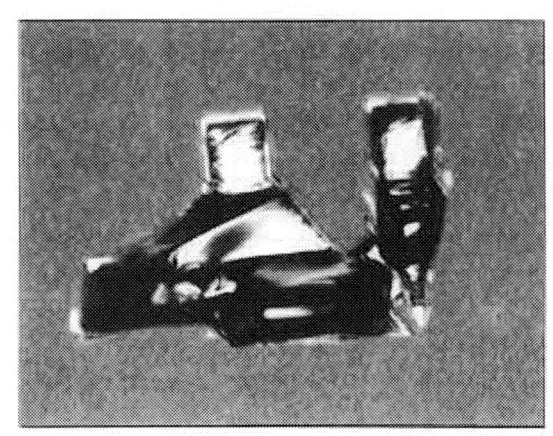

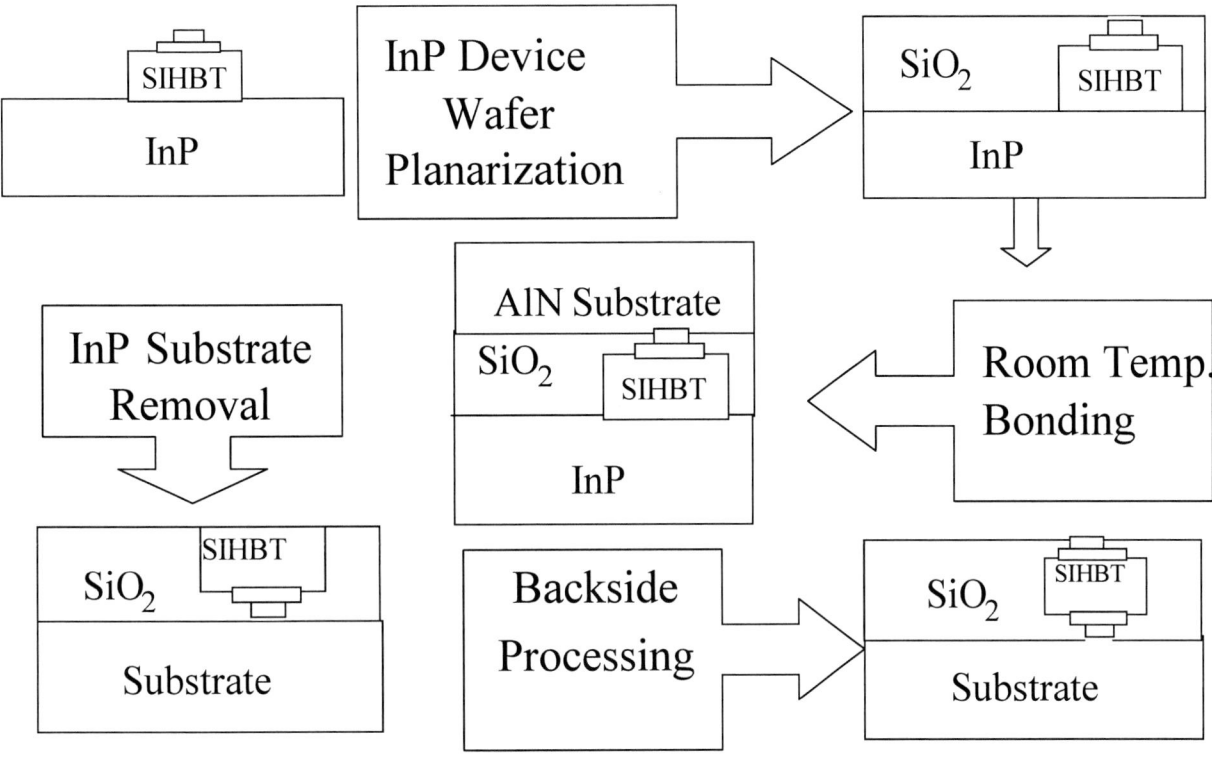

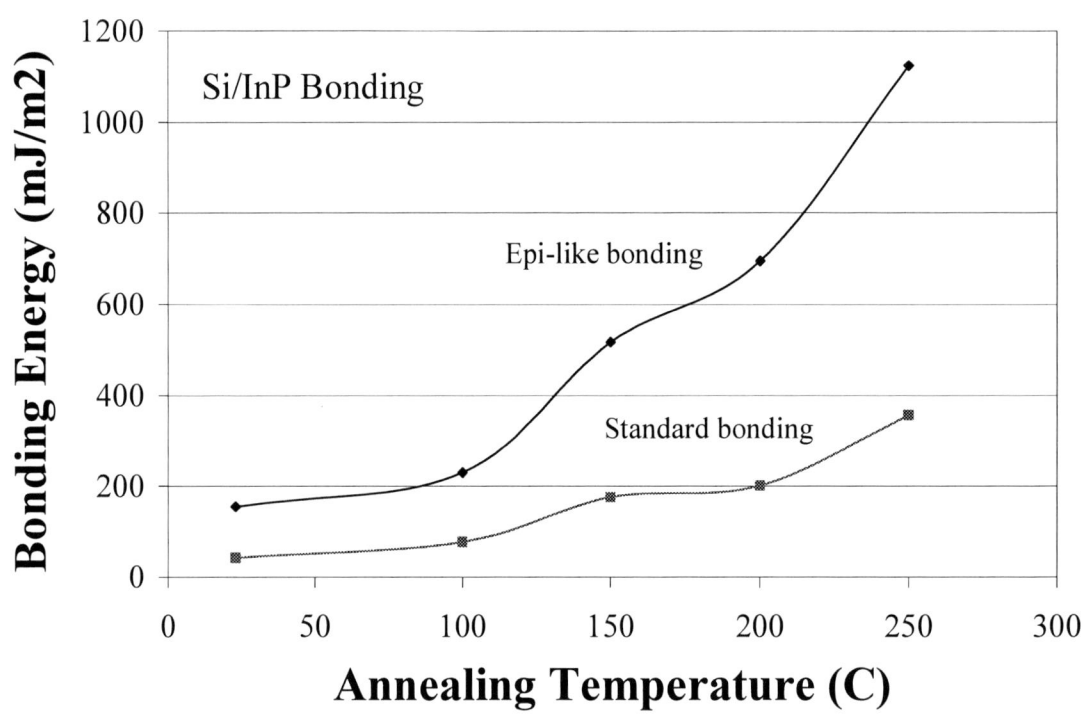

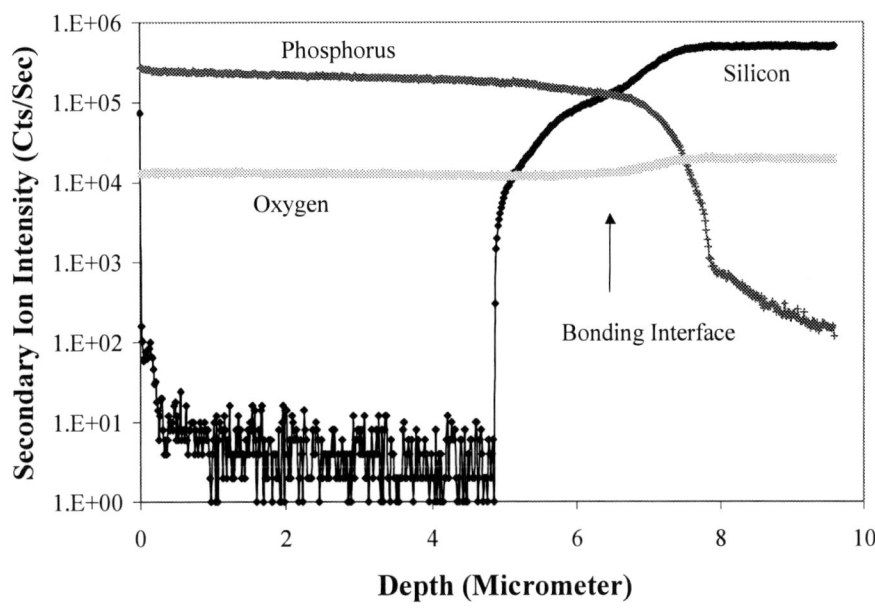

Mat. Res. Soc. Symp. Proc. Vol. 681E © 2001 Materials Research Society

Plasma Induced Chemical Changes at Silica Surfaces During Pre-Bonding Treatments

Darren M. Hansen[1], C.E. Albaugh[2], Peter D. Moran[1], and T. F. Kuech[1]

[1]Department of Chemical Engineering, University of Wisconsin-Madison, Madison, WI 53706, U.S.A.

[2]Department of Chemistry, Xavier University, Cincinnati, OH 45207, U.S.A.

ABSTRACT

Plasma-treated and DI H_2O rinsed oxide layers are commonly used in wafer bonding applications. Borosilicate glass (BSG) layers deposited by low-pressure chemical vapor deposition (LPCVD) treated with an O_2 plasma in reactive ion etching (RIE) mode at 0.6 W/cm^2 and rinsed with DI H_2O readily bond to GaAs and Si. The chemical role of this pre-bonding treatment was investigated using attenuated total reflection Fourier transform infrared (ATR-FTIR) spectroscopy. The peak intensities for both the Si-O and B-O absorbance bands decreased in intensity as a result the plasma treatment consistent with the uniform sputter etching of BSG. The effect of changing the total plasma treatment time was investigated in terms of the total amount of material removed. Polarization-dependent ATR-FTIR revealed that the H_2O/OH absorbance bands decreased in peak intensity with the OH groups preferentially oriented perpendicular to the sample surface after the plasma treatment. The subsequent DI H_2O rinse restores the water to the surface while changing the surface BSG composition. ATR-FTIR studies suggest that for oxide compositions greater than 10 mole % B_2O_3, the top 4 nm of B_2O_3 was removed or leached from the oxide layer during the DI H_2O rinse.

INTRODUCTION

Wafer bonding often employs oxide layers as versatile bonding media.[1] SiO_2 grown on Si substrates by thermal oxidation has primarily been employed in these applications.[1] Doped or multi-component oxides, however, provide greater flexibility in controlling the physical properties of the bonding medium. For example, alloys of B_2O_3 and SiO_2, forming a borosilicate glass (BSG), permit control over the viscosity of the oxide over many orders of magnitude at a given temperature.[2] The potentially low, controlled viscosity of BSG has been used in integrated circuit technologies to lower the thermal budget associated with planarization and reflow processes.

Independent of the oxide composition employed, strong bonding requires that the surfaces to be bonded be smooth, particle-free, and have surface H_2O/OH groups.[3] A hydrophilic oxide surface for wafer bonding can be prepared by either aqueous treatments or dry plasma treatments.[3-7] A common pre-bonding treatment that is being investigated as a means of forming strong room temperature bonds is exposure of the surface to an O_2 plasma followed by a DI H_2O rinse.[4-7] Previous studies have shown that bonding with GaAs and borosilicate glass (BSG)/GaAs is improved with this pre-bonding treatment.[8] The role of the pre-bonding O_2 plasma treatment has been suggested to include the introduction of a surface charge[4], the removal of hydrocarbons by etching of the surface[6], or the creation of a damaged oxide with unsatisfied surface SiO_x bonds that is said to be "activated"[5,7]. In this research, the goal is to probe the chemical changes induced by the O_2 plasma treatment and DI H_2O rinse using Fourier transform infrared spectroscopy (FTIR).

The goal of this study is to determine the chemical effect of the pre-bonding processing steps of O_2 plasma activation and DI H_2O rinsing on the BSG layer using attenuated total reflection Fourier transform infrared (ATR-FTIR) spectroscopy. First, changes in the chemical composition of the oxide layer were monitored by tracking the Si-O and B-O vibrations after each processing step. Different plasma exposure times and plasma power densities were investigated. In addition, the concentration of surface H_2O/OH groups was monitored since this is a critical issue for successful hydrophilic bonding.

EXPERIMENTAL PROCEDURE

Thin films of BSG were deposited in a conventional LPCVD reactor, that has been reported in detail elsewhere, at 675°C and 1 Torr total pressure.[12] Tetraethylorthosilicate (TEOS) and trimethylborate (TMB) were used as the SiO_2 and B_2O_3 sources, respectively. We have explored glass over a broad range of B_2O_3 concentration, but this report will focus primarily on 10 mole % B_2O_3 BSG. In this case, the growth rate is 6 nm/min and typical growth times were 5 minutes. Multi-wavelength spectral ellipsometry measurements, validated by profilometry measurements, were used to determine the initial BSG thickness as well as to confirm some results discovered by ATR-FTIR. The substrates for BSG deposition were 2.5 x 1.5 cm^2 semi-insulating GaAs substrates that were polished into trapezoids with 45° bevels for use in ATR-FTIR spectroscopy measurements.

The plasma treatment of the samples was performed in reactive ion etching (RIE) mode at 0.6 W/cm^2 in O_2 at a chamber pressure of 30 mTorr. Plasma-exposure time varied from 15 s to 15 min, with 5 min being a typical time. DI H_2O rinsing was performed for 5 minutes with ultrasonic agitation. Ultrasonic agitation is often used in the pre-bonding sequence to aid in the removal of void-inducing particles that may be trapped at the wafer-bonded interface.

Atomic force microscopy (AFM) was used to monitor the film morphology before and after plasma treatment, because surface roughness affects the nature of the bond.[3] AFM measurements were performed in contact mode using SiN tips over a 5x5 μm^2 area. The reported r.m.s. roughness values were obtained by averaging three values obtained from height scans on different regions of the BSG surface. Measurements of different samples at the same BSG composition were also performed to ensure consistent results.

ATR-FTIR measurements were performed with a resolution of 4 cm^{-1} over a spectral range of 650-1600 cm^{-1} and 3000-4000 cm^{-1}. The oxide related vibrations are all located between 650-1600 cm^{-1} for the BSG film.[13] The main SiO_2 stretching vibration is at 1078 cm^{-1} while the rocking vibration is at 801 cm^{-1}.[13] Both of these peaks can, therefore, be used to quantify the amount of SiO_2 present in the oxide layer. The main B-O stretching vibration is at 1380 cm^{-1} and can be used to quantify the amount of B-O present in the oxide layer.[13] One additional band, at 921 cm^{-1}, is also located in this spectral region and is attributed to a mixed B-O-Si vibration.[13] The spectral region between 3000-4000 cm^{-1} contains vibrations of H_2O/OH groups and can, therefore, be used to determine the amount of these species present on the surface.[9-10] All reported spectra are an average of 1000 scans that have been referenced to the GaAs sample prior to BSG deposition. The same sample was used for each step in the sequence ensuring the same sample volume for each spectrum. The triple phonon of GaAs at 770 cm^{-1} was used as an internal calibration standard.[11] Both p- and s-polarized scans of the sample were performed by introducing an IR polarization optic prior to IR beam incidence on the sample bevel in order to determine the orientation dependence of the surface species. At least three measurements were taken per sample and multiple samples were studied to ensure consistent and reproducible results.

RESULTS AND DISCUSSION

No change in surface morphology was detected on the scale accessible by AFM after any step in the pre-bonding sequence. Therefore, plasma exposure and DI H_2O rinsing do not adversely impact the criterion of smooth surface morphology for wafer bonding. A time lapse of ~24 hrs was required between plasma exposure and AFM investigation to obtain reliable data. A strong AFM tip-surface interaction prevented accurate measurements without the time delay.

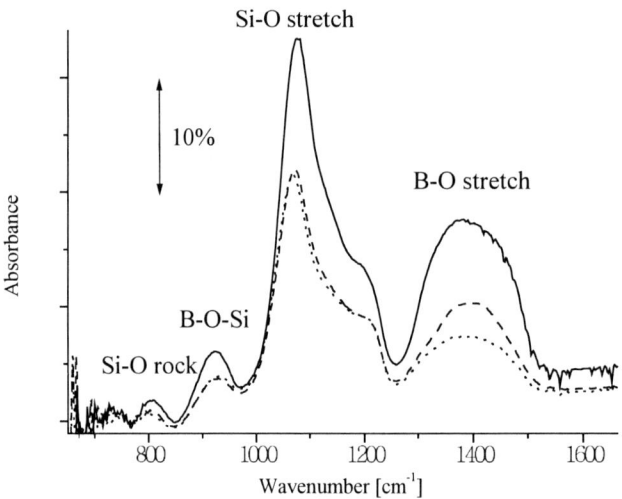

Figure 1. ATR-FTIR spectra of the oxide region for different treatments of a 10 mole % B_2O_3 BSG. The solid line (−) and the dashed line (---) represent spectra for the as deposited BSG and for the BSG after a 0.6 W/cm^2 O_2 plasma for 5 minutes, respectively. The decrease in peak intensity indicates the partial sputter removal of the BSG film. The dotted line (··) represents spectra after DI H_2O rinsing of the sample. The decrease in the B_2O_3 peak intensity indicates a surface removal or leaching of boron.

Figure 1 shows the ATR-FTIR spectra for the 10 mole % B_2O_3 BSG layer as-deposited, after the O_2 plasma treatment, and after the DI H_2O rinse. The as-deposited BSG-coated sample has peaks at 801, 921, 1078, and 1380 cm^{-1} within this spectral region, as expected. All of the peaks are still present, but at a decreased intensity after the plasma treatment. The SiO_2-related peak intensity decreased by 38% ± 2% and the B_2O_3-related peaks decreased by 40% ± 4%. The average and standard deviation were obtained from multiple samples that were all deposited to the same initial thickness. For each sample, p- and s-polarized ATR-FTIR spectra in addition to transmission FTIR measurements were made resulting in three separate FTIR measurements per sample. The Si-O to B-O peak intensity ratio was also used to determine changes in the average film composition. For the as-grown glass, the Si-O to B-O peak intensity ratio was 0.51 ± 0.05 while after the plasma treatment this value was 0.48 ± 0.05. The reduction in oxide peak intensity while maintaining a constant Si-O to B-O peak intensity ratio indicates a uniform removal of the BSG layer due to the plasma exposure. This decrease in film thickness was confirmed by spectral ellipsometry.

Figure 2 presents the difference in SiO_2 peak intensity between the as-deposited and plasma-treated BSG layers (I-I$_o$) as a function of plasma-exposure time. A linear fit to the data indicates a sputter etch rate of ~1.1 ± 0.2 nm/min for O_2 RIE plasma exposure at 0.6 W/cm^2. At 0.36 W/cm^2 a linear fit of similar data indicates a sputter etch rate of 0.32 ± 0.05 nm/min. Therefore, lower power plasmas also sputter the BSG layer, although at a slower rate. Lower power density plasma exposures were not investigated due to the long exposure times required to measure a change in the IR peak intensity.

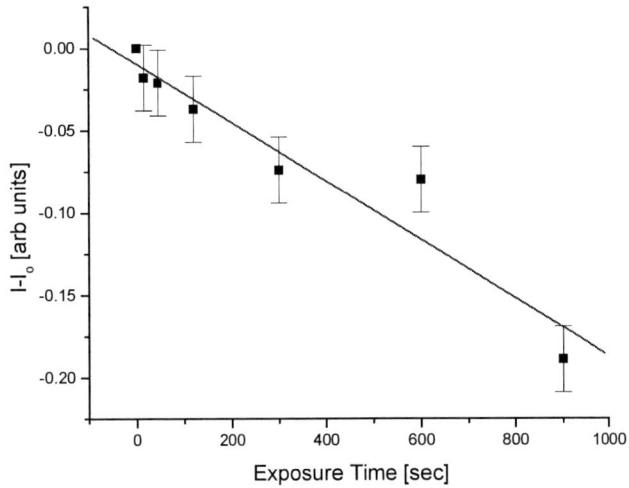

Figure 2: Time dependence for material removed during sputter-etching as a result of RIE plasma exposure. The difference between the SiO2 peak intensity before and after plasma treatment (ΔI, the amount of material removed) is plotted as a function of plasma exposure time in this figure. O$_2$ plasma at 0.6 W/cm^2 was used for these data.

After the DI H_2O rinse, only the B-O bands have decreased in peak intensity as indicated in Figure 1. The bands associated strictly with Si-O vibrations were unchanged after DI H_2O rinsing, indicating that the B_2O_3 component of the BSG is selectively removed from the surface as a result of the DI H_2O rinsing. The peak intensity of the B-O band decreased by 28% ± 4% for a 10 mole % B_2O_3 BSG layer. This result is consistent with the removal of B_2O_3 from the top 4.3 nm of the film. This effect was investigated for different BSG compositions. For 5 mole% B_2O_3 BSG, no selective removal of the boron was detected. 30 mole% B_2O_3 BSG exhibited a reduction in boron content corresponding to a removal of B_2O_3 from the top 5 nm of the layer, similar to the 10 mole% B_2O_3. For very high boron content BSG (>50 mole % B_2O_3) the entire oxide layer is water soluble, in agreement with previous reports.[13] This observation indicates that the glass composition near the surface where the bonding occurs has a different composition, and hence different physical properties, than the bulk oxide when 10 mole % B_2O_3 and higher BSG film compositions are rinsed with DI H_2O during the pre-bonding treatment.

Figure 3 presents the ATR-FTIR spectra for both p- and s-polarized spectra over the spectral range of 3000-4000 cm^{-1} before and after the O_2 plasma treatment. Figure 3(a) presents the spectra of p-polarized incident light while Figure 3(b) presents the spectra for s-polarized light. Figure 3(a) indicates that the plasma-treated sample has a reduced intensity of H_2O/OH compared to the as-deposited BSG layer. The reduced water-related peak intensity is attributed to the sputter removal of adsorbed water during the plasma treatment. Figure 3(b) shows the same general trend for s-polarization, however, the peak intensity is nearly extinguished after the plasma treatment. Since these two types of scans are expected to have equal sensitivity, affected only by the orientation of the surface species, these combined spectra indicate that the H_2O/OH groups on a plasma-treated BSG layer are predominantly oriented perpendicular rather than parallel to the plane of the BSG layer.[15] The plasma-treated surface is, therefore, a hydrophilic surface, still containing surface H_2O/OH groups.

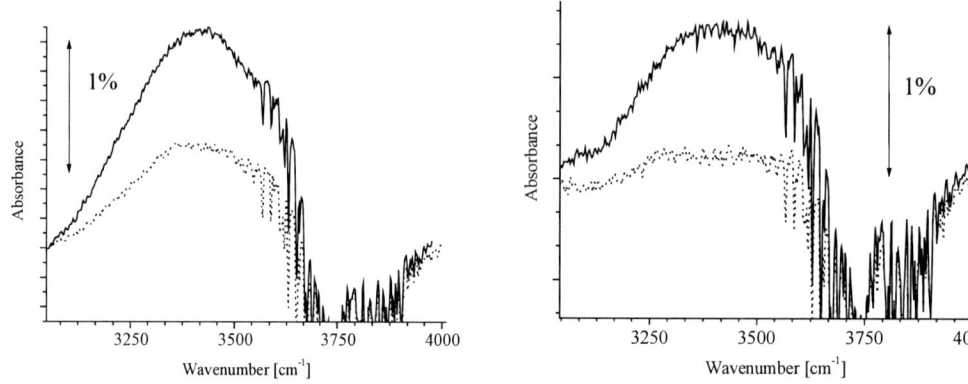

Figure 3. ATR-FTIR spectra for the H_2O/OH spectral region, 3000-4000 cm^{-1}. The p-polarized spectra are shown in (a) while the s-polarized spectra are shown in (b). In both cases, the solid line (−) and the dotted line ($^{...}$) represent spectra for the as-deposited BSG layer and for the BSG after a 0.6 W/cm^2 O_2 plasma treatment for 5 minutes, respectively. The H_2O/OH peak intensity is less after the plasma treatment, but the decrease is more pronounced for the s-polarized spectra indicating larger removal.

The spectra after a DI H_2O rinse, not shown in Figure 3, show an increase in intensity of the H_2O/OH-related peaks such that they slightly exceed the intensity measured prior to the plasma treatment for both the p- and s-polarizations. Both the p- and s-polarizations again have comparable intensity of H_2O/OH groups on the surface. ATR-FTIR is no longer able to detect an orientation preference for the H_2O/OH groups after the DI H_2O rinse treatment due to the large number of H_2O/OH groups present.

This study focused on the chemical changes detected by ATR-FTIR as a way of elucidating the role of the pre-bonding treatment. It is important to note that the pre-bonding sequence may affect other changes not detectable by FTIR measurements that facilitate the formation of a readily bondable surface. For example, it has been suggested that the plasma treatment may introduce a surface electronic charge that could aid the bonding process.[4] It is certainly possible that both the chemical and electrical roles of the plasma treatment combine to create an LPCVD BSG surface that is significantly easier to bond than the untreated oxide.

CONCLUSIONS

In summary, an O_2 plasma treatment and DI H_2O rinse of a BSG layer creates a surface that is much more readily bonded than the as-deposited BSG layer. ATR-FTIR studies indicate that the effect of the O_2 plasma is two-fold. O_2 plasma leads to the sputter removal of the BSG film at 1.1 ± 0.2 nm/min at 0.6 W/cm^2. At 0.36 W/cm^2, the sputter etch rate is 0.32 ± 0.05 nm/min. The sputter etching results in the removal of surface contaminants and the presentation of a fresh, chemically active surface. The plasma treatment also leads to the preferential perpendicular orientation of the H_2O/OH groups on the layer surface. This hydrophilic surface readily accepts additional H_2O/OH groups during the subsequent DI H_2O rinse. The DI H_2O rinse of the BSG layers removes or leaches B_2O_3 from the BSG surface for >10 mole % B_2O_3 BSG. These results indicate that the surface cleaning procedure prior to bonding determines the surface chemical composition through changes in the glass as well as the density of adsorbed H_2O/OH groups that facilitates the bonding process.

ACKNOWLEDGEMENTS

This work was supported by the ONR-MURI on compliant substrates at the University of Wisconsin-Madison. D.M. Hansen would like to acknowledge financial support from the Electrochemical Society through the summer 2000 Department of Energy Research Fellowship. Facilities support from the NSF UW-MRSEC is gratefully acknowledged.

REFERENCES

1. See references contained in *Semiconductor Wafer Bonding: Science, Technology, and Applications*, Volumes I-III, The Electrochemical Society Inc..
2. *Handbook of Glass Properties*, edited by N.P. Bansal and R.H. Doremus, Academic Press, New York, 1986.
3. Q.Y. Tong and U. Gosele, Materials Chemistry and Physics **37**, 101 (1994).
4. S.N. Farrens, J.R. Dekker, J.K. Smith, and B.E. Roberds, J. Electrochem. Soc. **142**, 3949 (1995).
5. V.H.C. Watt and R.W. Bower, Electron. Lett. **30**, 693 (1994).
6. D. Pasqariello, C. Hedlund, and K. Hjort, J. Electrochem. Soc. **147**, 2699, (2000).
7. M. Wiegand, M. Reiche, and U. Gosele, J. Electrochem. Soc. **147**, 2734 (2000).
8. D.M. Hansen, P.D. Moran, and T.F. Kuech, Mater. Res. Soc. Symp. Proc. **587**, O4.7.1 (2000).
9. Y.J. Chabal and S.B. Christman, Phys. Rev. B **29**, 6974 (1984).
10. M.K. Weldon, V.E. Marsico, Y.J. Chabal, D.R. Hamann, S.B. Christman, E.E. Chaban, Surface Science **368**, 163 (1996).
11. J.I. Pankove, *Optical Processes in Semiconductors* (Dover Publications, Inc., New York, 1971).
12. D.M. Hansen, P.D. Moran, K.A. Dunn, S.E. Babcock, R.J. Matyi, and T.F. Kuech, J. Cryst. Growth. **195**, 144 (1998).
13. E.A. Taft, J. Electrochem. Soc.: Solid State Science **118**, 1985 (1971).
14. D.S. Williams and E.A. Dein, J. Electrochem. Soc.:Solid-State Science and Technology **134**, 657 (1987).
15. Y.J. Chabal, Surface Science Reports **8**, 211 (1988).

Mat. Res. Soc. Symp. Proc. Vol. 681E © 2001 Materials Research Society

Molecular dynamics simulations of wafer bonding

Kurt Scheerschmidt

Max Planck Institute of Microstructure Physics, Weinberg 2,
D-06120 Halle/Saale, Germany, schee@mpi-halle.de
Tel: +49-345-5582910, Fax: +49-345-5582917

ABSTRACT

Molecular dynamics simulations using empirical potentials have been employed to describe atomic interactions at interfaces created by the macroscopic wafer bonding process. Investigating perfect or distorted surfaces of different semiconductor materials as well as of silica enables one to study the elementary processes and the resulting defects at the interfaces, and to characterize the ability of the potentials used. Twist rotation due to misalignment and bonding over steps influence strongly the bondability of larger areas. Empirical potentials developed by the bond order tight-binding approximation include π-bonds and yield enhanced interface structures, energies, and transferability to new materials systems.

INTRODUCTION

Wafer bonding, i.e. the creation of interfaces by joining two wafer surfaces, has become an attractive method for many practical applications to microelectronics, micromechanics or optoelectronics [1,2]. The macroscopic properties of bonded materials are mainly determined by the atomic processes at the interfaces during the transition from the adhesion state to the chemical bonding. Thus, the description of the atomic processes is of increasing interest to support the experimental investigations or to predict the bonding behaviour. While, in principle, it is now possible to predict material properties by using quantum-theoretical ab initio calculations with a miminum of free parameters, the only method to simulate atomic processes with macroscopic relevance is the molecular dynamics (MD) method using suitably fitted many-body empirical potentials. Such simulations enable a sufficiently large number of particles and relaxation times up to µs to be considered. However, the electronic structure and the nature of the covalent bonds can only be described indirectly. Therefore, it is of importance to find physically motivated semiempirical potentials starting mostly with the moments of the electron density and using tight-binding representations [3-5].

The MD simulations have successfully been used to describe ultra-high-vacuum bonding experiments for Si(100) [6], hydrogen passivated hydrophobic bonding processes [7], and to analyze the defect structure at bonded interfaces [8-10]. Simulations for SiC [11] were possible using the Tersoff [12] potential, whereas the predictions of the bondability of diamond have been performed using a bond-order potential [13]. Simulations of the bonding of amorphous silica (a-SiO_2) surfaces [14,15] may be the basis or a first step to describe hydrophilic wafer bonding. Conventional transmission (TEM) and high resolution electron microscopy (HREM) structure imaging has been applied to investigate the resulting interfaces and the defect structures at an atomic level [16], the MD relaxed structures are the basis to calculate TEM and HREM images from it, which in combination with calculated IR-spectra provides a good experimental evidence of the results.

METHOD

The method of molecular dynamics (MD) solves Newton's equations of motion for a molecular system, which results in trajectories for all particles considered in the system. The calculations are performed with a fifth-order predictor-corrector algorithm using a constant volume (NVE ensemble) or a constant pressure (NpT ensemble) and time steps of the order of 0.25 fs to ensure the proper calculation of surface modes. NVE is preferred for free surfaces and simulations to calculate diffusion processes, whereas NpT enables the relaxation of the cell dimensions and the application of an outer pressure, which is important for, e.g., the glass generation and the simulation of wafer interfaces. For controlling the system temperature the velocities are slightly rescaled at each time step of solely the atoms in the outer layers of the structure model, still applying periodic boundary conditions parallel to the interfaces, which describes an energy flux or dissipation into a macroscopic embedding substrate. In addition, for straight defects created at the interfaces the system is coupled elastically to the bulk wafers.

Simple pair potentials and potentials of the valence force field or related types are restricted in their validity to solely small deviations from the equilibrium. A better potential most often used for semiconductors is the Stillinger-Weber (SW) potential, consisting of 2- and 3-body interactions [17]. It allows the next neighbour interaction to be included by rescaling, which is a presupposition to the simulation of the dynamical behaviour without preordered surfaces and prescribed topology. Thus, e.g., the interaction of two silicon surfaces can be studied by correctly revealing the 2x1 reconstruction of a clean Si(100) surface. The potential used for silica is the modified Born-Mayer-Huggins (BMH) ionic pair interaction combined with a weak three-body term [18]. The BMH interaction combines a repulsion and a Coulomb term, which is screened to avoid the long-range Ewald sums, the three-body term is similar to SW. In addition, a Rahman-Stillinger-Lemberg (RSL) term was used for the water interactions.

Tersoff: empirical bond order potential

The potential of Tersoff [12] with different parametrizations TI-TIII has the shape of a bond order, which is a completely different functionality than SW, BMH etc:

$$V \sim \sum [\exp(-\lambda r_{ij}) - \mathbf{b_{ij}} \exp(-\mu r_{ij})]. \qquad (1)$$

The bonds are weighted by the bond order $\mathbf{b_{ij}}$ including all many body interactions over neighbours k different from the actual bonding pair i,j, and with all parameters fitted. It predicts the asymmetric reconstruction with fourfold coordinated atoms at Si interfaces with defects. Therefore it was applied to investigate the cores of 60° partial dislocations in Si and other defects left after bonding two wafers. Parametrizations exist also to describe the silicon-hydrogen interaction, hydrocarbons, SiC, Ge, etc. (see, e.g. [19]), thus hydrogenated Si(001) surfaces are well described including reconstructions of the (1x1) and (3x1) types [7] as well as Si-SiC interactions [11]. Because of the short range of the Tersoff potential it was supposed that the bond topology is given by the usual process starting with separated Si blocks of a suitable surface structure and orientation (surface reconstruction, steps) and applying long-range potentials initially.

Most of the existing potentials available are of the SW or T type. Compared to each other [20] they offer advantages and disadvantages in range of validity, physical meaning, fitting and accuracy as well as applicability. Such restrictions exist for other potential types, too, even if the (modified) embedded atom approximation is used (MEAM, [21]) or special environment dependencies are constructed to enhance the elastic properties near defects, as, e.g., in [22]. The interatomic forces in covalent solids, however, can only be completely described if the influence of the local environment according to the electronic structure is also included. A first step towards including such effects into an empirical description is the following bond order potential (BOP) approximation. In addition, no potential is applicable for long range interactions [23] which needs special consideration. The difficulty of developing suitable empirical potentials is threefold: One has to guess a functional form, motivated by physical intuition, fitted to ab initio and experimental data bases as well as reflecting nonfitted properties.

BOP: bond order approximation justified by TB methods

Tight binding approximations allow to develop physically motivated potentials, starting from analysis of the band energy:

$$E_{tot} = E_{rep} + E_{prom} + E_{band}(k) \tag{2}$$

$$\downarrow \qquad \downarrow \qquad \downarrow$$

$$\text{empirical} \quad s^2p^2\text{->}sp^3 \quad \sum U_{BO}\Theta_{BO} \approx b_{ij}=\pi_{ij}+(\sigma\pi_{ij}+\sigma\pi_{ji})/2 \tag{3}$$

$$\downarrow \qquad \downarrow$$

bond integrals   BOmatrix
Slater-Koster [24] Pettifor [5,25]
$ss\sigma$, $sp\sigma$,   Realspace GreenFkt
$pp\pi$...   Continued fraction

$$\downarrow \qquad \downarrow$$

BOP2=>Tersoff
BOP4=> π–bonds

The repulsive energy is assumed to be an embedded pair interaction, the promotion energy reflects the energy difference of valence s and p electrons, and the band energy yields the bond order term (cf. Eq. 3). The most important part is the expansion of the bond energy into hopping matrix elements, which are given directly by the two centre integrals of Slater and Koster [24], and bond order terms. The bond order terms b_{ij} can be approximated by Lanczo's recursion algorithm [25], where the level of the continued fraction determines the functional form and the applicability. A second moment approximation of the tight-binding model can be used to establish a general form at the level of the Tersoff potential with at least only four free fit parameters [4]. A further enhancement is possibly based on the bond order potential (BOP4) [5], which is given up to the fourth-level continued fraction of the Greens function. Its ability is demonstrated in the application to diamond wafer bonding [13].

RESULTS

The molecular dynamics simulation starting with two perfect and parallel-oriented Si blocks with perfectly aligned 2x1 reconstructed (100) surfaces and applying a slow heat transfer approach yields perfectly bonded structures [6,7]. However, a fast heat transfer, a starting configuration, with the dimer rows in orthogonal domain orientation, or including steps or small rotational misorientations, result in configurations no longer perfectly coordinated. The energy flux at surfaces is the driving force for the bonding process. The upper terraces behave like perfect surfaces, i.e., a weak attraction owing to the next neighbour interaction initiates the dimers to rearrange and to create new bonds. The energy the bonds have gained dissipates increasing the kinetic and elastic energies of the bulk. The resulting avalanche effect implies the bonding of the lower terraces, too. However, after bonding over double layer steps a disturbed interface and defects are left which may finally relax to 60° partial dislocations or shuffle-set dislocations accompanied by a row of vacancies [8-10]. Monolayer steps rotate the dimerization direction in the neighbouring domains and give rise to a stacking fault of either intrinsic or extrinsic type, depending on the dimer orientation in the adjacent terraces. Thus the interfaces between, or outside, the single-layer steps are characterized by a 90° twist-rotation, either with metastable fivefold coordination and Pmm(m) symmetry (SW potential or fast heat transfer) or (2x2) reconstructed and P(4)m2 symmetry (TSIII). The resulting local 42m dreidl configuration has been confirmed by DFT calculations and fits two rotated half crystals of minimal structural disorder and fourfold coordination, the interface energy is reduced by approximately 20%.

Twist rotation

The effect of a small twist angle as a rotational misorientation results in a mosaic-like interface structure. Figure 1 shows such interface defect structures after the MD simulation of rotationally misoriented bonding of (100) Si wafers and, for comparison, plan view TEM images of screw dislocation networks obtained in UHV bonding experiments of similar situations (Figures 1c,d). In Figure 1a) the waferbonding itself is simulated starting with two separated Si blocks having dimerized (100) surfaces and using the SW potential in MD. After bonding and sufficient relaxation under slow heat transfer conditions, almost all atoms have a bulk-like environment separated by misfit dislocations, which may have a high rate of kinks. In Figure 1b) the MD relaxation with a TIII potential is applied to a starting model with a prescribed network of two sets of a/2[110] screw dislocations accommodating a small rotational twist. The relaxed configuration shows symmetry breaking by twice the period of the array distance corresponding to the twist angle and having two different nodes T/S1 and T/S2. They are formed by symmetrical characteristic groups of atoms having the same point group symmetry 222 (D_2) as the core structures of individual screws. Most of the atoms forming T/S1 remain fourfold with large bond-angle and bond-length distortions, but there are two atoms in the unit with a fivefold surrounding. The T/S2-nodes are formed by more complicated atomic groups, however, showing solely a fourfold coordination. In Figure 1e) simulated plan view TEM contrasts based on these structures are shown assuming different thicknesses t and beam orientations [hkl] relative to the zone axis [001]. The orthogonal networks of the experiments (cf. Figs. 1c,d) possess a regular structure over large areas, distorted by steps due to the miscut of the wafers, holes, inclusions, or amorphized regions. Depending on the twist misorientation between the bonded wafers (usually 0.1° up to 3°), the annealing process after the bonding, and the imaging conditions (tilt, thickness) different details occur, which partially can be matched by the simulated images. MD simulations as well as TEM and HREM investigations showed that the screw dislocations forming the network of the (001) low-angle twist grain boundary can dissociate intrinsically into two 30° partials moving towards the bulk.

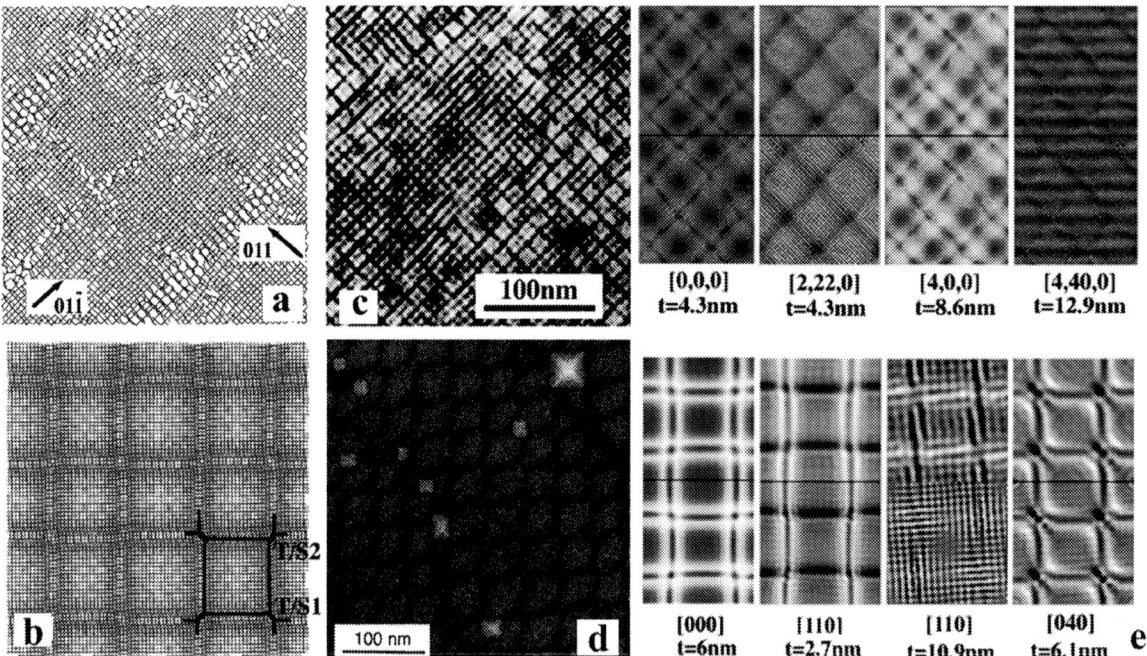

Figure 1. Rotationally misoriented waferbonding: a) Metastable kinked screw dislocation network of 4.6° rotated wafers after 2.5ps SW-MD relaxation at 900K, b) D=10b screw lattice (b=a/2<110>) at the TIII energy minimum showing two different nodes T/S1, T/S2, c,d) Experimental plan view images of screw dislocation networks of UHV-bonded (100) Si interfaces with different sample thicknesses and orientations (courtesy R. Scholz, MPI Halle), e) Simulated plan view TEM images for different thicknesses t and beam orientations indicated by the [hkl]-pole excitation (upper/lower row correspond to model a/b, resp.).

Using the Tersoff or BOP-like potentials and metastable or well-prepared starting configurations, yields further structure relaxation and energy minimization. Figure 2 shows some of the resulting minimum structures gained for higher annealing temperatures and different twist rotation angles (see Figure caption). A typical energy vs. time relaxation behaviour is shown in Figure 2a), at every temperature step - up and down - (selected are 4 of about 20 temperature steps) the system is relaxed for 25ps. Independently from the chosen twist angles and box dimensions all final structures yield finally bond energies of approximately 4.5 eV/atom at 0K. The energy gain, however, is directly related to the twist angle: approximately 0.02 eV, 0.03 eV, 0.05 eV, and 0.04 eV for 12.7°, 6.7°, 4.6°, and 2.8°, resp. The values are slightly modified if additionally steps or holes are included at the surfaces before bonding simulation starts. The maximum of energy gain between 4° and 6° twist is related to a change of the bonding behaviour itself: Whereas all simulations shown in Figure 2 with parallel dimerization at start clearly demonstrate the creation of the screw dislocation network, for orthogonal dimerization or small twist angles this is no longer valid.. As higher the annealing temperature as better the screw formation, i.e., similar to the prescribed perfect network and its node structure, cf. Figures 1b) and 2c,d). However, the detailed investigation of the energy gain and the influence of steps as shown in Figure 2e) in relation with simulated TEM and HREM and experiments as in [26] will be done in a forthcoming paper.

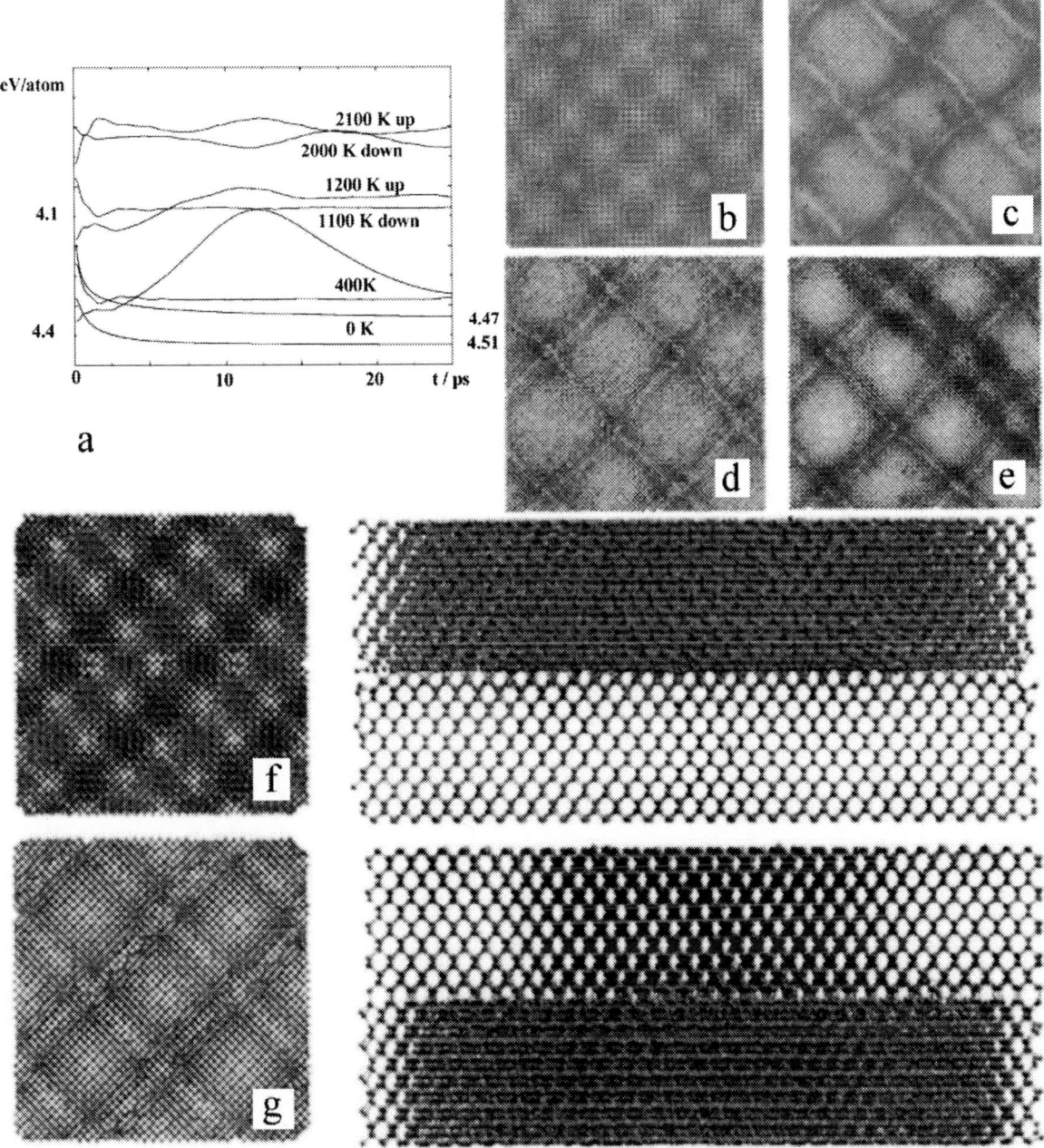

Figure 2. MD simulations of bonding rotationally twisted wafers with different angles and annealing temperatures: a) Typical energy relaxation during MD annealing cycles (up and down, three temperatures shown) starting and finnishing at 0K with -4.41 and –4.51 eV/atom, resp., b) Twist angle 2.8°, 134500 atoms, 22nm box, 2100K , orthogonal dimers, c) as b) parallel dimers, d) Twist angle 4.6°, 4.3nm box, 600K, e) as d) with [110]-steps, f) Twist angle 6.7°, 9.2nm box, 2100K , orthogonal dimers, [001] and [110] views, g) as f) with parallel dimers.

Adsorbates: H on Si and H_2O on Silica

The influence of adsorbates has been investigated, e.g., for hydrogenated surfaces assuming two Si(001)-3x1 blocks, corresponding to a hydrogen coverage of 4/3 monolayers, and NVE at 300K. A time step of 0.12 fs is used to account for the fast dynamics of the hydrogen atoms. External forces in the direction perpendicular to the interface are added to the interatomic forces of the two outermost atomic layers of each slab. The main result of the corresponding MD simulations consists in the finding of an energy barrier characterized by a critical pressure of about 80MPa to overcome the repulsion forces and to create covalent bonds across the hydrogenated interfaces. Figure 3a shows the oscillatory behaviour of the relative wafer distances below 80MPa, indicating the barrier, the bonding region between 100MPa and 1.4GPa, and the region of creating surface defects for pressures higher than approximately 1.4GPa.

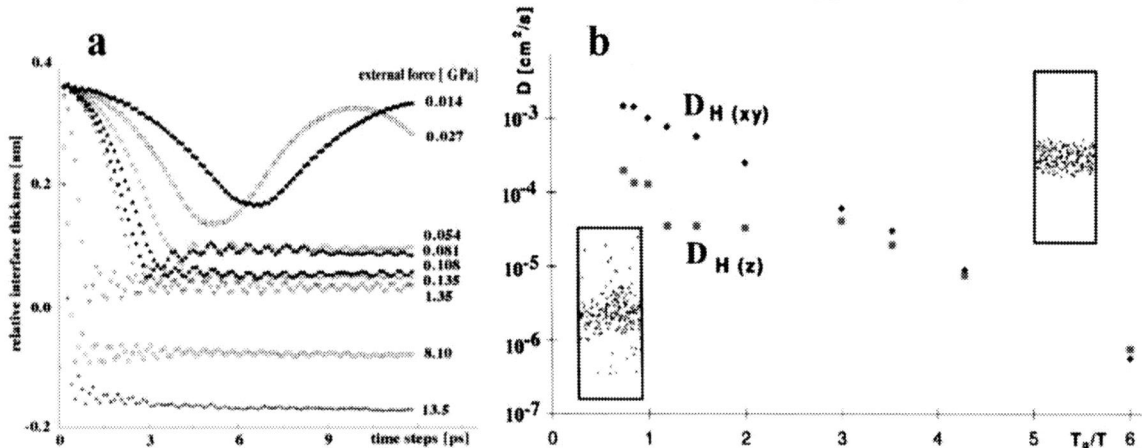

Figure 3. Energy barriers due to adsorbates: a) Transition between oscillatory behaviour (below 100MPa) and surface defect creation (above 1.4GPa) of Si wafer bonding with a hydrogen layer and applying pressure, b) Diffusion coefficients of hydrogen D_{Hxy} within and D_{Hz} across a bonded silica interface as function of the bond temparature relative to the glass transition temperature T_g of silica.

Figure 3b shows the hydrogen diffusion coefficients D_{Hxy} within and D_{Hz} across the interface as one of the main results of the MD simulation of silica amorphization, hydroxylation and bonding. The generation of amorphous silica models starts from crystalline silica structures by annealing cycles. Free surfaces were generated and the reconstructed surfaces were bombarded with H_2O groups. Such relaxed silica glasses have a highly reactive surface. Water molecules settling on the surface have at least three different kinds of bonding sites. They correlate to sites with reduced oxygen bridges, or where silanols are created by cracking Si-O bonds in the surface. Two adjoining hydroxylated surfaces, each covered with 1 to 2 monolayers of water are brought into contact to simulate the wafer bonding between silica and/or hydrophilic Si. Three different regimes are obtained, which can be related to experimental observations and the hydrogen diffusivity shown in Figure 2: The short-time behaviour at low temperatures and pressures leads to hydrogen-bonded surfaces with a low bonding energy; hydrogen has a high mobility in the gap. The bonding energy can be increased either by increasing the temperature and/or the pressure, or by tempering at lower temperatures for sufficiently long times. Increasing the temperature and/or pressure enables to dissolve the silanol and water groups and to lower the diffusion barrier, which occurs for temperatures between $T_g/3$ and T_g, with reduced hydrogen mobility across the gap. For higher temperatures the interface gap can be closed by forming direct silica-silica bonds, resulting in equivalent diffusivity parallel and normal to the interface. The long-time behaviour shows a reactive rearrangement of the surface, which locally leads to strong silica bonds even at low temperatures.

Diamond and SiC

The Tersoff potential is well suited to describe the SiC(0001)-3x3 and $\sqrt{3}x\sqrt{3}R30°$ surface reconstructions. A comparison with TB and DFT results shows that even the bond length and energy differences of the different reconstructions are correctly revealed. Applying similar simulations to the SiC(0001) wafer bonding yields for different starting configurations bond energies between 0.5 and 3.2 J/m^2. Figure 4a shows the relaxed final state for the $\sqrt{3}x\sqrt{3}R30°$ initial configuration resulting in a peak-to-peak arrangement. This configuration is the energetically optimized structure for which the prediction can be made that bonding in an ultrahigh vacuum environment should be possible.

Finally, in Figure 4b the BOP4 potential is used to simulate the bonding process of diamond. Two different surfaces are considered, viz. (001) and (111) with an equivalent reconstruction. Both the tight-binding MD and the semiempirical method using a BOP4 potential, which enables one to use far more atoms in the calculation and which demonstrates the quality of the fit, show the same bonding behaviour. The tight-binding level is necessary to describe correctly the π - bonds. The dimers remain, the π - bonds are broken during the bonding process, however, there is no graphitization at lower temperatures. The fourfold coordination is the resulting stable minimum structure. On the (111) surface, shown in Figure 4b, therefore the outermost dimers must be rearranged from threefold coordinations to fourfold ones.

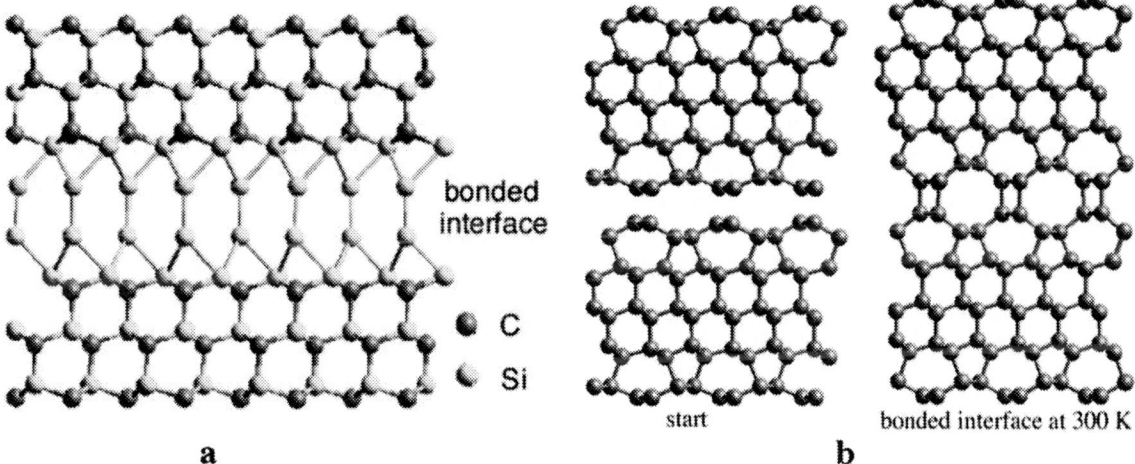

a   b

Figure 4. MD simulations of bonding SiC and diamond surfaces: a) SiC (0001)- $\sqrt{3}x\sqrt{3}R30°$ in peak-to-peak arrangement, b) C (111)-2x1 and using a BOP4 potential.

CONCLUSIONS

Molecular dynamics simulations (MD) based on empirical potentials are used to investigate the elementary steps of bonding two Si(001) wafers. The system is coupled elastically to the bulk wafers, and the energy dissipation is controlled by the transfer rates of the kinetic energy at the borders of the model. Calculated bonding energies and forces strongly depend on surface termination, native oxides, adsorbates, and process control. Twisted starting configurations, steps or rotational misorientations result in special interface configurations mostly no longer perfectly coordinated. Bonding energy and forces are determined not only by the surface structure but also by surface adsorbates as shown by MD of hydrogen on hydrophobic Si and by water-silanol reactions on silica surfaces describing the hydrophilic termination. The simulations lead to a better understanding of the physical processes at the interfaces and support the experimental investigations, especially the electron microscope structure analysis. Simulations based on potentials derived from the bond order expansion are used to enhance the physical reliability of the investigations and to predict the bonding behaviour of a wide variety of semiconductor surfaces.

REFERENCES

1. Q.-Y. Tong and U. Gösele, Semiconductor wafer bonding: Science and technology. Wiley, New York 1999.
2. A. Plößl and G. Kräuter, *Mater. Sci. Eng.* **R25** , 1 (1999).
3. C.M. Goringe, D.R. Bowler, and E. Hernandez, *Rep. Prog. Phys.* **60**, 1447 (1997).
4. D. Conrad and K. Scheerschmidt, *Phys. Rev.* **B58**, 4538 (1998).
5. D. G. Pettifor and I.I. Oleinik, *Phys. Rev.* **B59**, 8487 (1999).
6. D. Conrad, K. Scheerschmidt, and U. Gösele, *Appl. Phys.* **A62**, 7 (1996).
7. D. Conrad, K. Scheerschmidt, and U. Gösele, *Appl. Phys. Lett.* **71**, 2307 (1997).
8. A.Y. Belov, D. Conrad, K. Scheerschmidt, and U. Gösele, *Philos. Magazine* **A77,** 55 (1998).
9. A.Y. Belov, K. Scheerschmidt, and U. Gösele, *phys. status solidi* **171**, 159 (1999).
10. A.Y. Belov, R. Scholz, and K. Scheerschmidt, *Philos. Mag. Lett.* **79**, 531 (1999).
11. C. Koitzsch, D. Conrad, K. Scheerschmidt, and U. Gösele, *J. Appl. Phys.* **88**, 7104 (2000).
12. J. Tersoff, *Phys. Rev.* **B38**, 9902 and **B39**, 5566 (1989).
13. D. Conrad, K. Scheerschmidt, and U. Gösele, *Appl. Phys. Lett.* **71**, 49 (2000).
14. S. H. Garofalini, *Electrochem. Soc. Proc.* **93-29**, 57 (1994).
15. D. Timpel, M. Schaible, K. Scheerschmidt, *Journ. Appl. Phys.* **85**, 2627 (1999).
16. K. Scheerschmidt, D. Conrad, A. Belov, H. Stenzel, *Electrochem. Soc. Proc.* **97-36**, 381 (1998).
17. F.H. Stillinger and T.A. Weber, *Phys. Rev.* **B31**, 5262 (1985).
18. S. H. Garofalini, *J. Non-Cryst. Solids* **120**, 1 (1990).
19. A.J. Dyson and P.V. Smith, *Surface Science* **355**, 140 (1996).
20. S. Balamane, T. Halicioglu, W.A. Tiller, *Phys. Rev.* **B46,** 2250 (1992).
21. M.I. Baskes, *Phys. Rev.* **B46**, 2727 (1992).
22. J. F. Justo, M.Z. Bazant, E. Kaxiras, V.V. Bulatov, S. Yip, *Phys. Rev.* **B58,** 2539 (1998).
23. Y. C. Wang, K. Scheerschmidt, and U. Gösele, *Phys. Rev.* B **61**, 12864 (2000).
24. J.C. Slater and G.F. Koster, *Phys. Rev.* **94**, 1498 (1954).
25. D.G. Pettifor, *Phys. Rev.* **B63**, 2480 (1989).
26. J.L. Rouviere, K. Rousseau, F. Fournel, and H. Moriceau, *Appl. Phys. Lett.* **77**, 1135 (2000).

Mat. Res. Soc. Symp. Proc. Vol. 681E © 2001 Materials Research Society

Wafer Bonding of Silicon Carbide and Gallium Nitride

Jaeseob Lee, T. E. Cook, E. N. Bryan[1], J. D. Hartman, R. F. Davis, and R. J. Nemanich[1, 2]
Department of Material Science & Engineering and Department of Physics[1]
North Carolina State University,
Raleigh, NC 27695
[2]E-mail address: Robert_Nemanich@ncsu.edu

ABSTRACT

Wafer bonding of SiC and GaN may prove important in the formation of high power heterojunction devices. Results of bonding SiC (C or Si surface) onto GaN (Ga surface) are presented. The samples were n-type 6H SiC and epitaxial n-type 2H(wurzite) GaN grown on SiC. The results demonstrate bonding for both possibilities, but the bonding of the C surface SiC to Ga surface GaN is more readily accomplished. A lower resistance was found for the C-face SiC/Ga-face GaN. The results indicate that the polarity of the interface is important for bonding of these materials.

INTRODUCTION

Silicon carbide is considered for high power and high temperature semiconductor device operation because the material exhibits a wide bandgap (3.0 eV) and high thermal conductivity (5W/cm°C). The operation of silicon carbide bipolar junction transistors with an even larger bandgap emitter would display increased current gain due to improved emitter efficiency. The larger bandgap of the emitter would restrict the diffusion of holes from the base to the emitter, resulting in high electron injection efficiency into the base. Additionally, the increased bandgap of the emitter allows the base to be heavily doped, thereby decreasing the base resistance. Also, since SiC is an indirect bandgap material, free carriers exhibit longer lifetimes, compared to a direct bandgap material such as GaN. The increased lifetime yields a long diffusion length, and a high base transport factor. Furthermore, these devices have a short base width, which further enhances the transport factor, thereby increasing the current gain. Gallium nitride is a natural choice for a larger bandgap emitter for SiC. Gallium nitride not only has a higher bandgap, 3.4 eV, than SiC, it also has a high thermal conductivity, 1.3W/cm°C. With a lattice constant of 3.18Å for GaN and 3.08Å for SiC epitaxy is possible, but the lattice mismatch (~3.4%) can significantly limit the quality of the epitaxial film [1,2].

The poor wetting of GaN on 6H-SiC(0001) substrates impedes direct nucleation and frequently results in GaN films of poor crystallinity. The use of an AlN buffer layer has been demonstrated to be effective in improving the crystallinity as well as in reducing the defect density in the GaN films, but simultaneously inhibits carrier injection across the AlN/SiC interface due to the insulating nature of AlN and its high band gap [3]. Reports of the growth of GaN directly onto 6H-SiC(0001) have noted the observation of an amorphous interlayer or zinc-blende inclusions at or near the GaN/SiC interface. Epitaxial growth by conventional techniques, e.g., molecular beam epitaxy (MBE) and metal organic chemical vapor deposition (MOCVD), is affected by lattice mismatch strain. During MBE and MOCVD growth dislocations form to relax the strain energy as the layer thickness exceeds a critical value, known as the critical thickness. It is energetically preferred for dislocations to nucleate on the surface of the layer, subsequently gliding down towards the interface, and drawing threading dislocations behind, most of which remain during further layer growth [4]. The critical thickness is on the order of 0.2µm. The dislocation density observed for GaN on SiC at a thickness of 0.5µm is typically about 5×10^9 cm^{-2} [3].

Direct wafer bonding is an alternative method of forming a heterointerface or heterojunction. It is an appropriate technique for materials of high lattice mismatch or with chemical instabilities. The method of direct bonding enables the formation of atomic bonds across atomically flat surfaces of different materials without introducing threading dislocations. Direct wafer bonding has proved its capability in enabling new classes of heterogeneous devices [4]. This study explores approaches for wafer bonding of SiC and GaN. Mechanisms of direct bonding are proposed based on the experimental observations. We also report initial I-V measurements of a SiC/GaN bonded interface.

EXPERIMENTAL DETAILS

The SiC wafers were obtained from Cree Research, and 50 mm 6H (0001)$_{Si}$ n-SiC and 50 mm 6H (000$\bar{1}$)$_C$ n-SiC wafers used. A 1.5µm thick 2H(wurzite) (0001)$_{Ga}$ n-GaN film was grown by MOCVD onto (0001)$_{Si}$ SiC with a (conducting) AlN buffer layer. The (0001)$_{Si}$ SiC wafer and the (0001)$_{Ga}$ GaN film were on-axis, but the (000$\bar{1}$)$_C$ SiC wafer was cut 3° off-axis. Doping levels (n-type) were 2~4$\times 10^{18}$ cm^{-3} in the SiC wafers and 1×10^{16}~10^{17} cm^{-3} in the GaN film. The SiC wafer showed a resistance of less than 0.1 ohm-cm. The wafers were diced into 10$\times$10 mm^2 pieces and subjected to a sequential bonding process consisting of *ex situ* wet cleaning, *in situ* dry cleaning, *ex situ* bonding and *in situ* annealing. *Ex situ* wet cleaning includes a standard RCA SC1 clean (NH$_4$OH: H$_2$O$_2$: H$_2$O=0.25:1:5, 80°C, 10min), DI water rinse (3min), N$_2$ blow dry and a standard RCA SC2 clean (HCl: H$_2$O$_2$: H$_2$O=1:2:8, 80°C, 10min), DI water rinse (3min), N$_2$ blow dry. A final HCl dip was employed for the GaN and a final HF dip was used for SiC substrates (each 1min). *In situ* dry cleaning includes ~800°C 10min UHV annealing of SiC and ~550°C 10min UHV annealing of GaN.

After dry cleaning samples were removed from the UHV chamber into air and (0001)$_{Si}$ SiC/(0001)$_{Ga}$ GaN and (000$\bar{1}$)$_C$ SiC/(0001)$_{Ga}$ GaN pairs were positioned in a chuck and loaded back into UHV. Bonding was initiated through pressure applied to the surfaces while at room temperature. After the initial bonding, *in situ* annealing was employed to form a high density of bonds at the heterointerface. The *in situ* UHV annealing process included up to 5 annealing cycles at 890°C for times up to 30min. Longer anneal times were employed for the (0001)$_{Si}$ SiC/(0001)$_{Ga}$ GaN couples while a 10min anneal was sufficient for bonding of (000$\bar{1}$)$_C$ SiC/(0001)$_{Ga}$ GaN couples. The surfaces were characterized prior to bonding with AFM and AES. After the bonding sequence, IR light images of the bonded pairs were recorded to characterize the degree of bonding. I-V measurements of the bonded pair were completed after deposition of 2000Å of Ti on both sides of the bonded pair.

RESULTS

As is typical for as-received SiC wafers, the surfaces showed polishing scratch grooves, and 20x20 µm^2 AFM scans displayed an RMS roughness of 20±5Å. The GaN surfaces displayed similar grooves, which are also apparently related to the polishing scratches of the initial SiC substrates. An AFM image of a SiC surface is shown in Figure 1. The chemical properties of the surfaces were examined by AES. To obtain measurements representative of the surfaces prior to bonding, the samples were removed from vacuum and held in ambient air for a period of time typical for bonding. The samples were then reloaded into the UHV system for AES measurements. The AES indicated monolayer oxidation of the SiC surface and the presence of monolayer levels of oxide and carbon contamination on the GaN surfaces (Figure 2).

After *in situ* cleaning the surfaces were installed in a special sample holder to initiate the bonding process. The bonding process involved two steps, initial contact bonding with the samples placed in the holder in ambient air, and annealing of the contact bonded couple to initiate covalent bonding. Identical experiments were carried out for $(0001)_{Si}$ SiC/$(0001)_{Ga}$ GaN and $(000\bar{1})_C$ SiC/$(0001)_{Ga}$ GaN. Many attempts to form a bonded interface using annealing temperatures less than 800°C were unsuccessful. For annealing at ~900°C bonding was obtained for both SiC surfaces. For the $(0001)_{Si}$ SiC/$(0001)_{Ga}$ GaN partial bonding was obtained in only one experiment out of five attempts. However, for $(000\bar{1})_C$ SiC/$(0001)_{Ga}$ GaN effective bonding occurred in each of two attempts. For the latter case a single ten minute anneal proved effective while anneals for as long as 30 min. were attempted for the Si surface of SiC.

After the bonding process, the couple was removed from UHV and the bonding was characterized optically using infrared light. The presence of interference fringes indicates incomplete bonding. In addition, pressure applied to the surface will cause shifting of the fringes of an incomplete bonded surface. Optical images showed that the $(0001)_{Si}$ SiC/$(0001)_{Ga}$ GaN pair displayed partial bonding while the $(000\bar{1})_C$ SiC/$(0001)_{Ga}$ GaN pair displayed bonding over the whole area of the couple (Figure 3). Shown in Figure 3 are three couples after the bonding process. One $(0001)_{Si}$ SiC/$(0001)_{Ga}$ GaN was annealed to ~900°C but bonding was not observed. In another case, where five 900°C, 30 min. annealing cycles were employed, partial bonding was observed. For the $(000\bar{1})_C$ SiC/$(0001)_{Ga}$ GaN pair, a uniform bonded interface was observed on each of two attempts.

The electrical properties of the bonded couples were examined after metallization of the back side of the SiC wafers. Noting that both samples were n-type, we may expect near ohmic behavior. I-V measurements showed that the $(000\bar{1})_C$ SiC/$(0001)_{Ga}$ GaN pair has a smaller resistance than $(0001)_{Si}$ SiC/$(0001)_{Ga}$ GaN pair (Figure 4). The lower resistance indicates a more effective interface bond.

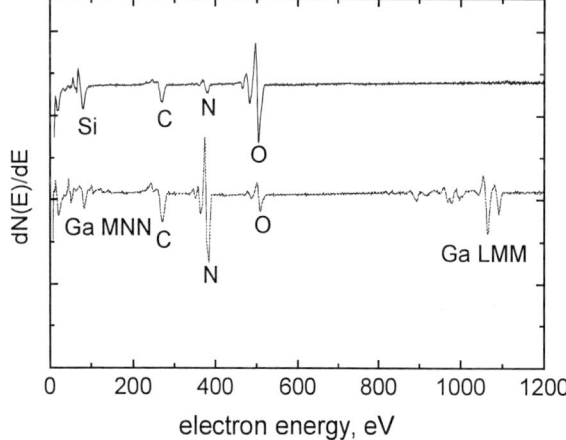

Figure 1. AFM image of as-received SiC show RMS roughness 20 ± 5 Å in 20×20 μm^2 area

Figure 2. AES analysis of SiC and GaN after reloading from the air

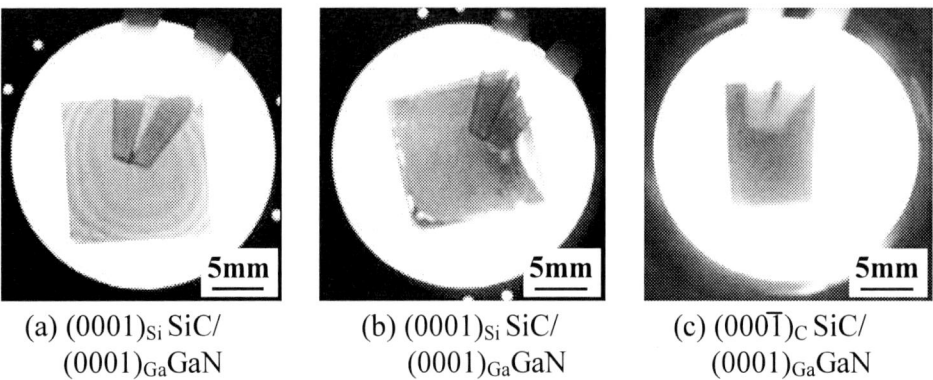

| (a) $(0001)_{Si}$ SiC/ $(0001)_{Ga}$GaN | (b) $(0001)_{Si}$ SiC/ $(0001)_{Ga}$GaN | (c) $(000\overline{1})_{C}$ SiC/ $(0001)_{Ga}$GaN |

Figure 3. IR image of pairs after (a)UHV annealing of 890°C 30min, (b) 5 cycles of UHV annealing between RT and 890°C 30min, (c) UHV annealing of 890°C 10min

DISCUSSION

The results presented here indicate a significant difference in the effectiveness of bonding the C or Si surface of SiC to GaN. In both cases, annealing to ~900°C was necessary. It is likely that the annealing assists the bonding through contaminant removal and through atomic displacements to optimize the formation of chemical bonding.

It has been reported that in the case of epitaxial growth of GaN on SiC the strength of covalent bonds is diminished by the polarization induced by the surrounding atoms and the ionic component of the bonding is significant [5]. Therefore, the lowest energy interface occurs when "positive" ions bind to "negative" ions. This polarity determination can be applied to wafer bonding. With knowledge of electronegativity values, we can assume the bonding process of the elements (Figure 5). Table 1 gives the electronegativity values of the respective elements of GaN and SiC. Table 2 shows the difference in electronegativity and the amount of partial ionic character in the possible bonds [6].

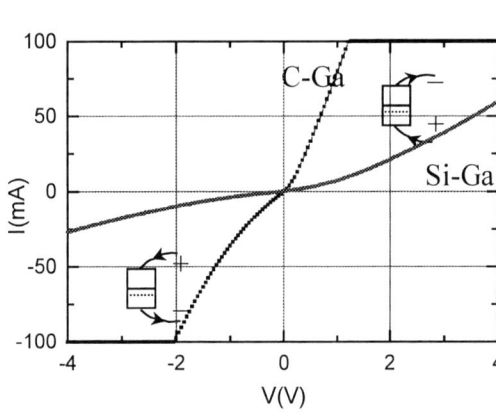

Figure 4. I-V of directly bonded Si face SiC/Ga face GaN pair and C face SiC/Ga face GaN pair

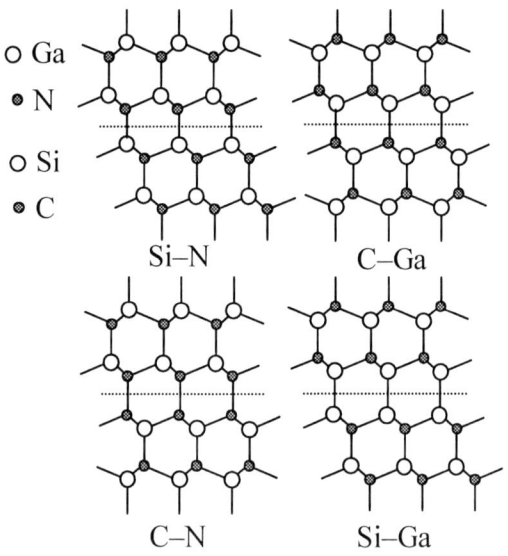

Figure 5. Schematic [11$\overline{2}$0] projection of the GaN/SiC interface

Table 1. Electronegativity values	
Element	Electronegativity
N	3.0
C	2.5
Si	1.8
Ga	1.6

Table 2. Bond and ionic character		
Bond	Difference in Electronegativity	Ionic Character
N-Ga	1.4	39%
N-Si	1.2	30%
C-Ga	0.9	19%
C-Si	0.7	12%
N-C	0.5	7%
Si-Ga	0.2	1%

Of the four possible interface bonds, the Si-N bond has the highest electronegativity difference and the largest ionic character. So Si-N bonds at the interface will form a strong bond, which is partly covalent and partly ionic. The C-Ga bond has a slightly lower electonegativity difference and still has a 19% ionic character. So the C-Ga bonds will also provide a strong covalent and ionic bond. In contrast, N-C and Si-Ga have small ionic components, which will result in weaker bonding.

With this viewpoint of electronegativity difference, our results of effective bonding for $(000\bar{1})_C$ SiC/$(0001)_{Ga}$ GaN and only partial bonding of $(0001)_{Si}$ SiC/$(0001)_{Ga}$ GaN are explained.

CONCLUSIONS

The results presented here demonstrate the potential of forming a bonded interface between GaN and SiC surfaces. Polarity is an important factor in wafer bonding of these materials. Ga-terminated GaN readily bonds to C-terminated SiC, but it was difficult to bond Si-terminated SiC to GaN. The $(000\bar{1})_C$ SiC/$(0001)_{Ga}$ GaN pair has the lower interface resistance. We are proposing that heterojunction devices [1,2] may be prepared from directly bonding $(000\bar{1})_N$ n-type GaN and $(0001)_{Si}$ p-type SiC. We are now exploring approaches to obtain smooth surfaces and to minimize surface contamination.

ACKNOWLEDGEMENTS

This research was supported by the Office of Naval Research, MURI (Multidisciplinary University Research Initiative) project, Project No. 98PR05894-00, Award No. N00014-98-1-0654 (John Zolper, monitor).

REFERENCES

1. J. Pankove, S.S. Chang, H.C. Lee, R.J. Molnar, T.D. Moustakas, and B. van Zeghbroeck, *Int. Electron. Dev. Meet. Tech. Dig.*, pp. 389-392 (1994).
2. S.S. Chang, J. Pankove, M. Leksono, and V. van Zeghboeck, *Dev. Res. Conf. Dig.* pp. 106-107 (1995).
3. B. Yang, A. Trampert, B. Jenichen, O. Brandt, and K.H. Poog, *Appl. Phys. Lett.*, **73**(26), 3869, 28 Dec. (1998).

4. N.Y. Jin-Phillipp, W. Sigle, A. Black, D. Babic, J.E. Bowers, E.L. Hu, and M. Ruhle, *J. Appl. Phys.*, **89**(2), 1017, 15 Jan. (2001).
5. R.B. Capaz, H. Lim, and J.D. Joannopoulos, *Phys. Rev. B.*, **51**(24), 17755, 15 Jun. (1995).
6. L. Pauling, *The Nature of the Chemical Bond*, 3rd ed., Cornell university Press, Ithaca, NY, p.93 (1960).

Mat. Res. Soc. Symp. Proc. Vol. 681E © 2001 Materials Research Society

TEM measurement of hydrogen pressure within a platelet

J. Grisolia, G. Ben Assayag, B. de Mauduit, A. Claverie
CEMES/CNRS, BP 4347, F-31055 Toulouse
R.E. Kroon, J.H. Neethling,
Physics Department, PO Box 1600, University of Port Elizabeth, South Africa

ABSTRACT

Proton implantation and thermal annealing of silicon result in the formation of a specific type of extended defect involving hydrogen, named "platelets". These defects have been related to the exfoliation mechanism on which a newly developed process to transfer thin films of silicon onto various substrates is based. In a previous paper, we have shown that these platelets undergo a quasi-conservative Ostwald ripening upon annealing. The measurement of the pressure within such pressurised gas-filled cavities is important to understand and simulate both the growth of these defects and the exfoliation mechanism. To extract this pressure from TEM studies, we have developed and tested an analogy between the platelets and a well-known 2D defect: a dislocation loop. The comparison between simulations of the image of the strain field surrounding a fictitious dislocation loop and experimental TEM images of the platelets shows that the platelets can be described by a Burgers vector of about 0.6nm. Moreover, this vector can be used to deduce the pressure of the molecular hydrogen within a platelet. A typical value of 10 GPa is found for a platelet of 20 nm in diameter at room temperature. Consequently, the atomic density of hydrogen within a platelet and the total number of hydrogen trapped by a population of platelet can be calculated and give reasonable values when compared to the implanted dose.

INTRODUCTION

Extended defects formed after high dose hydrogen implantation have been intensively studied during the last years due to their potential technological applications in advanced silicon processing. More specifically, H-rich two dimensional (2D) cavities, the so-called platelets, have recently received considerable attention because they are used in the microelectronic industry to obtain the delamination of a thin film from a thick substrate [1,2].

In a previous paper, we have shown that, after a sufficiently high dose H implantation and low temperature annealing, nucleation and growth of 2D platelets occurs [3]. These platelets are thought to be 2D precipitates of hydrogen atoms bounded either to Si atoms, to vacancies or to other H atoms. During annealing, these defects grow in size and reduce their density with kinetics that has allowed us to identify a conservative Ostwald ripening mechanism. Thus, during annealing these defects exchange H atoms. The driving force for this mechanism is the reduction of the formation energy (the energy cost to add one extra atom to the defect) consecutive to the size increase of the platelets [4]. In the meantime, while the overall elastic energy decreases in the implanted layer, the strain locally increases around the projected range of the protons i.e., where the platelets tend to concentrate and where splitting will ultimately occur.

Establishing the link between the growth of platelets at the microscopical level with the macroscopically observed layer splitting or fracture of the material requires the knowledge of the stress field created by such objects. Moreover, the modelling of the Ostwald ripening of these platelets is in principle possible but the three master equations which describes the phenomenon heavily depend on the value of the hydrogen pressure inside the platelets. To measure this strain field experimentally and deduce the hydrogen pressure, we have carried out strain contrast analysis around platelets with the TEM. We have tested a geometrical analogy between a platelet and a dislocation loop of unknown Burgers vector. By comparing computer simulated and experimental images obtained under specific contrast conditions, this Burgers vector has been obtained so that the hydrogen pressure within a platelet of about 20 nm in diameter could be deduced.

EXPERIMENTAL DETAILS

To form platelets, p-type (boron doped $\sim 5 \times 10^{14}$ cm^{-3}) (100) CZ silicon wafers were implanted at room temperature with a dose of 3×10^{16}H$^+$.cm^{-2} at an energy of 61keV and subsequently annealed under N$_2$ at 500°C for 1h30. Cross-section samples were prepared following the standard procedure involving mechanical thinning and ion beam milling at about 100 K until electron transparency. To measure their size distribution, platelets are often imaged under "out-of-Bragg" strongly defocused conditions [3, 4]. Under such conditions the strain field which surrounds the defect cannot be imaged. For this purpose, other imaging conditions have to be used.

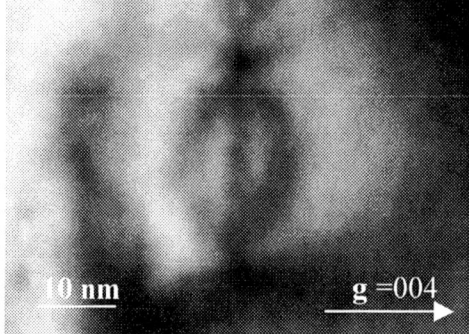

Fig. 1 : TEM image of a platelet on a(001) plane viewed edge-on (B = [110]), g = 004, s=0).

Figure 1 is a Bright-Field (BF) image of one "typical" platelet taken under two-beam conditions when the Bragg condition is exactly satisfied for **g** = 004 at 200 keV. This platelet is viewed edge-on and lies on the (001) plane. Its diameter is of about 23 nm. Since the defect appears in the middle of the first dark thickness fringe, the local thickness of the foil can be estimated to be of about half the extinction distance. Moreover, this defect is located half-way from both free surfaces, in the middle of the foil, as it can be deduced from the symmetry of the image.

ANALOGY BETWEEN PLATELETS AND DISLOCATION LOOPS

It is noticeable that under most imaging conditions platelets appear with contrast which reminds us of the strain contrast characteristic of dislocation loops. Indeed, two lobes are observed symmetrically to the habit plane of the defect and are due to the extension of the strain field in the vicinity of the defect. This observation is quite consistent with the finding resulting from Raman experiments that the (100) platelets are filled with hydrogen mostly in the molecular state [5]. This gas under pressure tends to bend the silicon lattice similarly to what is observed in the case of an extrinsic dislocation loop

A dislocation loop is totally defined by its Burgers vector. Indeed, the displacement field surrounding a defect of a given size can be calculated from this single vector. If we assume the platelet to generate a strain field "somehow" similar to that created by a dislocation loop, it is possible to define a pseudo or an "effective" Burgers vector of a platelet. Moreover, the amplitude of the displacement field perpendicular to the habit plane of the defect is necessarily related to the internal pressure of the hydrogen gas.

This effective Burgers vector can be defined as being the mean swelling of two (100) planes surrounding the center of the defect. Figure 2a represents a fictitious intrinsic (vacancy type) dislocation loop lying on a (100) plane and having a Burgers vector equal to 0. Fig. 2b represents a platelet the volume of which can be written $V_f = \pi r^2 (d_{100} + b_{eff})$ where d_{100} is the distance between the (100) planes and b_{eff} is the effective Burgers vector of the platelet.

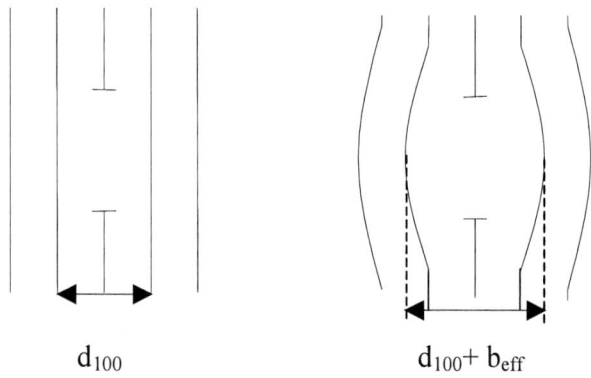

$$d_{100} \qquad\qquad d_{100}+ b_{eff}$$

Figure 2 : Geometrical representation of a) an intrinsic dislocation loop and

b) a H₂ filled platelet.

IMAGE SIMULATIONS

Image simulation techniques can be used to determine this Burgers vector through a best fitting procedure. For this, the comparison with images taken under specific experimental conditions such as shown by Fig. 1 have to be used. Under such conditions, the dynamical two beam theory can be applied under the so-called column approximation. The numerical values that we have used to run the simulations are summarised in the following table :

Parameter	Value
Beam direction B	[1 1 0]
Loop normal n	[0 0 1]
Diffraction vector g	0 0 4
|g|	0.737 nm
Extinction distance ξ_g	160 nm
Foil thickness T	0.50 ξ_g
Loop depth in foil D	0.25 ξ_g

Loop radius r	Initially 12 nm, also 10 nm
Effective Burgers vector magnitude |b|	0.4, 0.5 and 0.6 nm
Deviation from the Bragg condition w	0
Normal absorption coefficient N	0.026
Anomalous absorption coefficient A	0.026
Poissons ratio v	0.222

Table 1 : Numerical values used for the simulations [6, 7, 8, 9]

Fig. 3 is a set of simulations corresponding to dislocation loops of 20 nm in diameter. At the bottom, the experimental image shown in Fig. 1 is also reported at the same magnification for comparison. The 3 simulations in the left column correspond to dislocation loops of increasing Burgers vector (b).

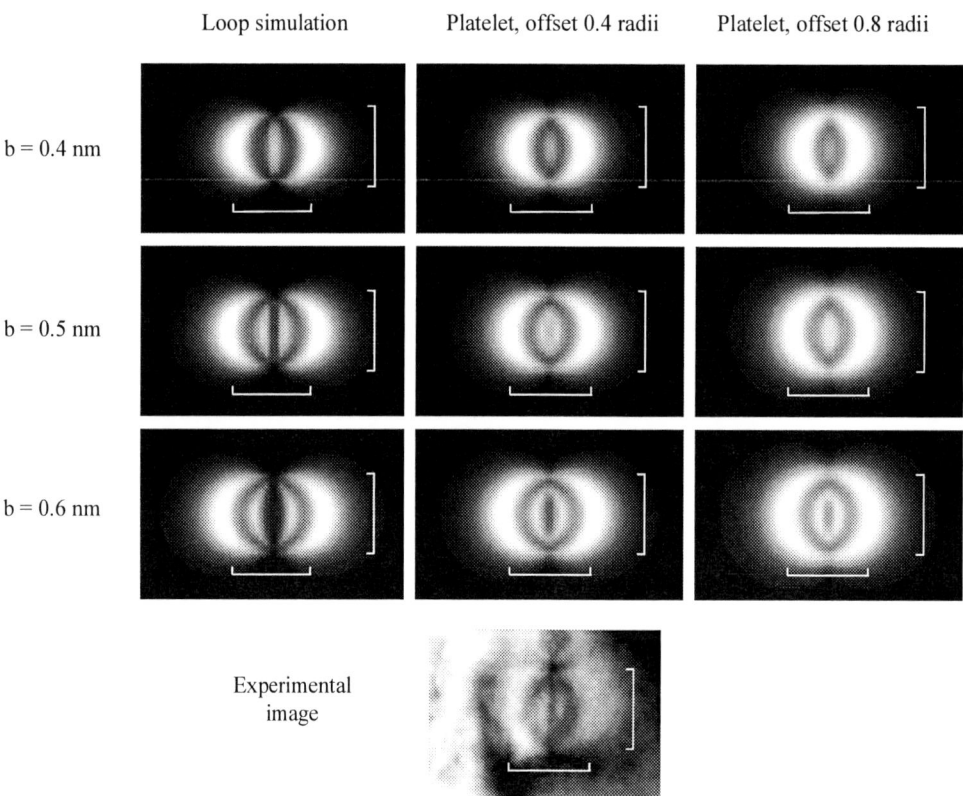

Fig. 3 : Set of TEM image simulations of dislocation loops and platelets

While the general characteristics of the experimental image are found for the simulation ran with b=0.5 nm, the image of a platelet is clearly more "round shaped" than of a real dislocation loop. This evidences the limitation of the analogy we have developed. However, to obtain a better match, an offset distance which artificially shifts the displacement field towards the habit plane of the defect was introduced in the simulations. Under such conditions, an excellent fit could be obtained using b=0.6 nm and an offset distance of about 4 nm. This shows that, whatever is the exact structure of the defect, the displacement field created by a platelet of about 20 nm in diameter is quite similar to the one created by a dislocation loop of the same diameter and having a Burgers vector in the 0.5-0.6 nm range. While it would be a very large value for a "real" dislocation loop (usually b is in the 0.1-0.3 nm range), this result is consistent with values found by others on nearly identical objects [10-15].

GAS PRESSURE WITHIN A PLATELET

It is very tempting to try to deduce the gas pressure within a platelet from the knowledge of the displacement field surrounding it and assuming again than the analogy with a dislocation loop holds. If we consider the equilibrium of platelet of radius r, volume V, containing n molecules, an incremental increase of the volume dV generates a work PdV, where P is the pressure of the gas. On the other hand, the elastic energy of a dislocation loop can written,

$$W_{el} = \frac{rb_{eff}^2\mu}{2(1-v)}[\ln(\frac{8r}{\alpha})-1],$$ where μ, v, α are respectively the shear modulus, the

Poisson ratio and α a constant.

The energy increase of the platelet due to a small increase of b_{eff} must be equal to the work PdV. Thus, we obtain,

$$P = \frac{b_{eff}\mu}{\pi r(1-v)}[\ln(\frac{8r}{\alpha})-1],$$

which for a radius of 10 nm and b=0.6 nm gives P in the 10 GPa range.

This relation shows that the pressure is proportional to the Burgers vector we derive from strain contrast analysis. This pressure decreases when the diameter of the platelet increases. Since our analysis have been performed at room temperature, it is expected that the pressure and the associated strain fields are (much) larger during the Ostwald ripening of the platelets which typically takes place at temperatures in the 350-500°C range.

To estimate the volumic density of the gas within such platelets, one can use the expression of Kamada [6] that relates the hydrogen molar volume (cm³/mol) to the gas pressure. Under such assumptions, one can find that the gas density is almost independent on the diameter of the platelet but is again proportional to b_{eff}. For b_{eff} =0.6 nm, one gets a density of about 1×10^{23} mol/cm³.

To test the validity of our calculations, one can use these numerical values to estimate the number of hydrogen atoms $N_{hydrogen}$ (cm⁻²) trapped by a population of platelet. Indeed, we have,

$$N_{hydrogen} = d_{at} \cdot \pi (<r>^2 + \sigma^2) \cdot e \cdot d$$

where d_{at} is the gas atomic density, $<r>$ is the mean radius of the population, σ the standard deviation, e the thickness of the platelet and d the surface density of the population. Using the numerical values we have previously obtained for the size and density evolution upon annealing we find that about half of the initially implanted dose is retained within the platelets in good agreement with the SIMS analysis we have already performed [7]. This last result shows that the different numerical values we have generated under the assumption that platelets resemble dislocation loops appear realistic.

REFERENCES

1- M. Bruel, Electron. Lett., 31 (14), (1995), 1201.

2- B. Aspar, C. Lagahe, H. Moriceau, A. Soubie, M. Bruel, A.J. Auberton-Hervé, T. Barge and C. Maleville, Mat. Res. Soc. Symp. Proc., V510, (1998), 381, .

3- J. Grisolia, L. Laanab, B. Aspar, C. Lagahe, G. Ben Assayag, A. Claverie, Appl. Phys. Lett., Vol 76 (7), (2000), 852.

4-- J. Grisolia, F. Cristiano, B. De Mauduit, and G. Ben Assayag, F. Letertre, B. Aspar, L. Di Cioccio, A. Claverie, Journal of Appl. Phys, 87 (12) (2000) 8415.

5- B.Aspar, C. Lagahe, E. Jalaguier, A. Soubie, B. Biasse, A.M. Papon, J. Grisolia, A. Claverie, T. Barge, F. Letertre. IIT Proceedings, 2000, in press.

6- G. Radi, Acta. Cryst. (1970), A26, 41.

7- R. E. Kroon, J. H Neethling, S. Afr. J. Sci., 95, (1999) 349.

8- J. P. Hirth, J. Lothe, Theory of dislocations, McGraw-Hill, New York (1968).

9- O. Madelung,, Semiconductors-Basic Data, Springer-Verlag, Berlin (1996).

10- S.P. Ardren ., C.A.B Ball., J.H. Neethling and H.C. Snyman, J. Appl. Phys. 60, (1986), 695

11- J. R. Botha, Kroon, R.E. and Neethling, J.H., Proc. Microsc. Soc. South Afr. 27, (1997)

12- R.E. Kroon. and J.H. Neethling, *Proc. 14th Int. Cong. on Electron Microscopy* (1998), 51

13- S. Muto, S. Takeda, M. Hirata, T. Tanabe, J. Appl. Phys.70, (1991) 3505.

14- S. Muto, S. Takeda, Phil. Mag., 72, (1995) 99.

15- S. Muto, S. Takeda, M. Hirata, Material Science Forum 1171, ((1995) 196.

16-J. Grisolia, F. Cristiano, G. BenAssayag, A.Claverie, Nucl. Instr. And Meth. in Phys. Res. B (2001), in press.

Mat. Res. Soc. Symp. Proc. Vol. 681E © 2001 Materials Research Society

Blistering on Silicon Surface Caused by Gettering of Hydrogen on Post-Implantation Defects

A.Y. Usenko[1], and W.N. Carr[2]
[1]Silicon Wafer Technologies Inc., 240 King Blvd., Newark, NJ, 07102 USA
[2]New Jersey Institute of Technology, 323 King Blvd., Newark, NJ, 07102 USA

ABSTRACT

A well-known process for thinning of silicon by slicing submicron-thick crystalline films from substrates uses direct implantation of protons. In this paper we describe a different way of delivering hydrogen to a cleavage plane. In our process a defect-rich buried layer is first formed with ion implantation. Defects in the as-implanted silicon work as traps for hydrogen. Next monatomic hydrogen is delivered to the trap layer by electrolytic charging. To check sliceability, the samples were annealed and blistering was observed. Evidence of blistering is a sign of potential cleavage. The electrolytic charging was performed using a simple two-electrode cell. The front side of the as-implanted silicon wafer was exposed to an electrolyte. The backside of the wafer was contacted with an aluminum layer and connected to a current source. The acidic electrolyte was buffered with ethylene glycol. Buffering was used to suppress bubbling on the wafer surface and to improve the uniformity of charging. To increase charging current the wafer was illuminated with visible light. A graphite rod was used as the positive electrode in the cell. A few Coulombs per square centimeter of the wafer were passed through the cell during the hydrogenation process. The depth of blisters is about 1/2 of projection range of the implanted ions. It means that the hydrogen platelets are formed in the region of maximum of vacancy-enriched post-implantation defects. This process of electrolytic hydrogen charging may be used in future to manufacture silicon-on-insulator wafers with very thin top silicon layer. Thin SOI offers important advantages in the production of substrates for mainstream CMOS integrated circuit manufacturing.

INTRODUCTION

Wafer bonding combined with delamination of crystalline films from single crystal substrates is used in a well-known silicon-on-insulator technology [1] called Smart-Cut™. In the Smart-Cut process, to delaminate the silicon film, hydrogen ions are implanted with a high dose (more than 4×10^{16} cm^{-2} for H^+, and $2 \cdot 10^{16}$ cm^{-2} for H_2^+) into the silicon wafer [2]. In this process a dense hydrogen-rich layer is formed followed by heating to promote defect transformation. The implanted hydrogen precipitates into platelets or microbubbles with heating. Depending on how deep the hydrogen layer is located, the precipitation results in either (a) cleavage of silicon along the hydrogen layer or (b) blistering of silicon surface. If the hydrogen layer is deeper than few microns, or if the surface is stiffened with an additional bonded or deposited layer, then cleavage proceeds. If hydrogen precipitates closer than about a micron from a surface, the film blisters before the continuous layer formation. Cleavage and blistering phenomena are caused by the same hydrogen precipitation mechanism. The cleavage is a useful phenomenon, while blistering is an undesirable phenomenon. Nevertheless, evidence of blistering is a sign of potential cleavage.

A drawback of the Smart-Cut process is that a heavy implant dose (at least 2×10^{16} cm^{-2}) is required. An attempt to decrease the dose ("smarter cut") [3] by Goesele at al. uses boron-then-hydrogen sequential implantation. They found experimentally that the threshold dose for blistering decreases if hydrogen is implanted after boron, and without annealing between the implantations. However, the total dose drop with the "smarter cut" is not significant. Other research groups [4,5] confirmed independently that co-implantation results in a relatively small decrease in total dose needed.

We conclude that the threshold dose for blistering in the "smarter cut" is caused by trapping of hydrogen onto defects created by the boron implantation. This leads to our process in which a thin hydrogen layer in silicon is formed using electrolytic hydrogenation instead of direct implantation.

EXPERIMENTAL DETAILS

This report uses 100 mm n-type silicon wafers, phosphorus doped, 0.01 to 4 Ohm cm, <100> orientation, 525 μm thick, single-side polished implanted with silicon at 180 keV, 5×10^{15} cm^{-2}. Wafers were kept at –25°C during implantation. Then the as-implanted samples were immersed into a 2-electrode electrolytic cell compatible with a cleanroom environment [6]. For high resistivity wafers we used the cell with the backside contact of Fig.1. For low resistivity wafers we used a simpler cell with the wafer edge contact of Fig.2. Since the cell in [6] was used for anodic oxidation, we reversed the polarity, and some other minor changes were made. A 100-mA current source was connected to the cell containing mixture of ethylene glycol and H_2SO_4 as an electrolyte. An integrated charge of 1 to 10 C/cm^2 was passed through wafer during the process. Charging starts at room temperature. After reaching 0.01 Coulomb the temperature was increased to 100°C allowing higher current. After electrolytic charging the wafers were cleaned. Then the wafers were annealed at 550°C. The wafer surfaces were then analyzed with optical microscope.

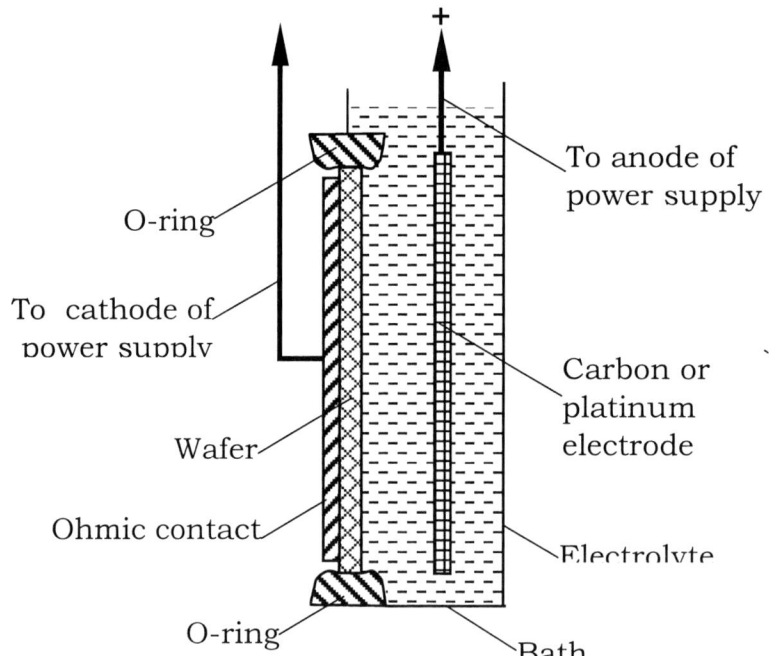

Fig.1. Electrolytic cell for hydrogenation of silicon wafers

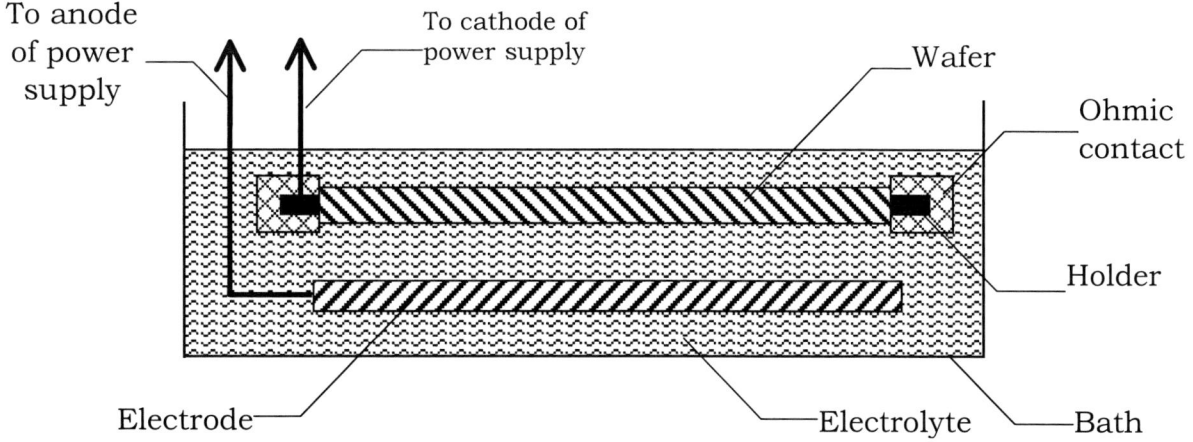

Fig. 2. Simplified electrochemical cells used for hydrogenation of low resistivity wafers.

RESULTS

Fig.3 is an optical microscopy view at 500× of a wafer processed with the self-implantation, hydrogenation, and followed by annealing. The surface is covered with features with lateral dimensions about 1 micron with partial, full, and broken blisters. The wafer surface of Fig.4 is obtained by the proton implant [2] (60 keV, 10^{17} cm^{-2}) with annealing at the same temperature 550°C, but without electrolytic hydrogenation. We find the surfaces of Fig.3 and Fig.4 are indistinguishable in feature types. Control wafers (as-implanted, implanted-and-annealed without electrolytic charging, etc.) do not show blistering.

DISCUSSION

Unique features of hydrogen in silicon supports formation of the buried hydrogen layer:

High diffusivity	atomic form is mobile even at room temperature
Trapping efficiency	attaches to defects (broken bonds, dislocations, impurities, boundaries)
Ease of cluster formation	hydrogen-rich layer transforms into layer of platelets and bubbles

The high diffusivity (~10^{-10} cm^2/s at room temperature) of monatomic hydrogen in crystalline silicon has been known since classic measurements were performed by Van Wieringen and Warmoltz 50 years ago. To diffuse monatomic hydrogen into silicon several techniques can be used. In this paper we report electrolytic charging of a silicon wafer [7-10].

To accumulate hydrogen in the desired volume of the wafer we "pre-form" defects that readily interact with hydrogen. Implantation is our process to create buried defect layers. Each implanted ion produces up to 10^4 atomic displacements, and up to 10^3 secondary vacancy-containing defects that are potential traps for hydrogen. Implantation at lowered temperature suppresses defect self-annealing and results in higher concentration of post-implantation defects. Silicon self-implantation has the advantage of not contaminating or doping the wafer.

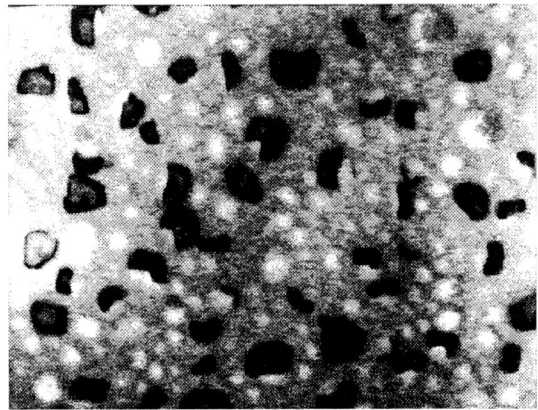 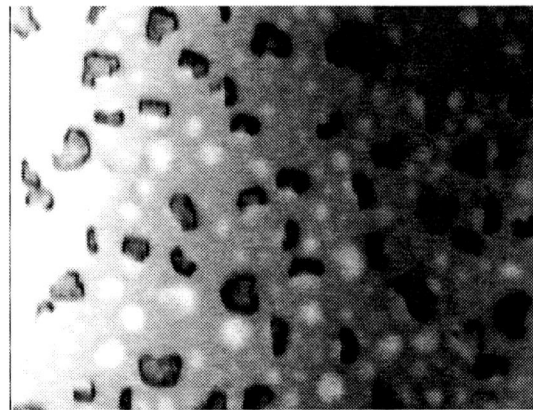

Fig.3.Blistered surface of self-implanted silicon wafer after electrolytic charging with hydrogen (x500).

Fig.4. Blistered surface of silicon wafer after high dose hydrogen implantation (x500).

The above listed unique features are mostly related to protons and to atomic hydrogen and much less related to molecular hydrogen due to differences in their sizes and chemical activities. Therefore, we supply hydrogen in atomic form. Hydrogen plasma, and liquids that dissociate releasing hydrogen (acids, water) can be considered as hydrogen singlet sources. A properly directed electric field will help to deliver hydrogen inside of the wafer. The electrolytic hydrogenation we followed in this paper uses techniques described by de Mierry [7], Pearton [8], Oehrlein [9], and Raghavan [10].

Hydrogen trapping is an important step of the process we suggest. Trapping of hydrogen onto point, line, and extended defects in crystalline silicon have been investigated quite extensively [11-13]. However, mechanisms of hydrogen capture onto densely spaced traps are still not well understood. Recent theoretical evaluations by Cerofolini, et al. [14,15] show that there is no strong decrease in capture efficiency with densely packed traps. Blistering that we experimentally observe is consistent with their model. Nickel at al. [16] found that hydrogen platelets can be created under the silicon surface by rf plasma hydrogenation of initial device-quality silicon. They found that a critical process step is the platelet nucleation at temperatures lower than 250°C. Platelets can then be grown efficiently at higher temperature. These results are also consistent with our results.

The implant dose needed to form an effective trap layer for hydrogen is in the range of 10^{13} to 10^{15} cm^{-2}. This dose is still much lower than the dose needed for conventional processes. During and after the implant process many of the displaced atoms return to their previous places in the lattice. Thus many of the displacements self-anneal or annihilate. In practice, 90 to 99% of the displacements do annihilate, depending mostly on the target temperature. At low temperature (liquid nitrogen) self-annealing is suppressed, and at higher temperature the self-annealing effect increases. That is why low temperature implantation is preferred [17,18]. Upper level of the dose is limited by amorphization of surface. We used 5×10^{15} cm^{-2} (sub-upper) dose in our experiments.

When a bombarding ion penetrates the silicon lattice, the displacement cascade proceeds in a cigar-shape volume oriented along the ion trajectory. The displacement cascade is also known as the disordered region. The disordered region contains an inner vacancy-rich region and an outer interstitial-rich region. The vacancy-rich region serves as an effective hydrogen trap while the interstitial-rich region does not trap the hydrogen effectively. Monatomic hydrogen diffuses through the interstitial-rich outer region to become trapped in the inner part of the displacement cascade. The interstitial-rich layer has a moderate (~0.7 eV) potential barrier for hydrogen. The diffusion coefficient of hydrogen through outer part of the displacement region is low. This diffusion coefficient is about equal to diffusion coefficient of hydrogen in amorphous silicon. The diffusion coefficient of monatomic hydrogen H^+ in crystalline silicon is $9.7 \times 10^{-3} \exp(-0.48 eV/kT)$ from Van Wieringen and Warmolz. The diffusion coefficient for monatomic hydrogen in the interstitial-rich region is about 4 orders of magnitude lower.

A further increase in the implantation dose increases the thickness of the amorphized layer, and the upper boundary of the amorphized layer at some high dose level reaches the silicon surface. Widening of the amorphized layer is undesirable, because to effectively lift-off a layer the fracture plane should be defined precisely and thin. Otherwise the platelets appear on different crystal planes causing an increased surface roughness of the delaminated layer. An amorphized layer of minimum thickness contains about 10^{18} cm^{-2} hydrogen trap sites which is enough for further continuous platelet layer formation. This minimum thickness is approximately 0.01 micrometer and is approximately equal to the size of inner (vacancy-rich) region of the disordered region. Going beyond the amorphization threshold is not desirable. There is an optimum implant density which results in near-atomically flat surface of the delaminated layer.

An explanation of the low efficiency of sequential implantation based processes ("smarter cut" process [3] and others [4,5]) may be as follows. Hydrogen implantation produces defects and the as-implanted wafer contains the hydrogen in a trapped form. In the subsequent annealing process Goesele teaches [3] that the implanted hydrogen releases from its attachments to hydrogen implantation-induced defects, and attaches to boron implant-induced defects. Releasing from attachments requires annealing out of defects. Both hydrogen- and boron-implantation-induced defects have a similar microscopic nature and similar annealing temperatures. Therefore when hydrogen is released, the boron-induced traps are mostly annealed out, and the hydrogen re-attachment process is not very effective. That is why only a 20% improvement in total implant dose was obtained in [3].

CONCLUSION:

Buried defect-rich layer formation and subsequent decoration of these the defects with hydrogen can cause silicon surface blistering. The same process is suggested to cut thin single crystalline layers from a substrate. The defects work as a trap. A buried amorphized layer can be used as the trap. The implant dose is expected to be an order of magnitude lower compared with the dose in a conventional Smart-Cut™ process. Different techniques, including electrolytic charging or plasma can be used to fill the trap with hydrogen. The process may be used to fabricate low-cost SOI wafers.

ACKNOWLEDGEMENTS

The authors gratefully acknowledge Dr. Dentcho Ivanov from New Jersey Institute of Technology for his help with wafer processing, and Mr. R. Sinclair of Implant Sciences for help with edge-to-edge wafer implantation, and low-temperature implantation.

REFERENCES

1. A.J. Auberton-Herve, "Commercialization Of Thick And Thin SOI by The Smart Cut™ Process, in this Proceeding," (2001).

2. M.Bruel, *Electronics Letters*, **31**, 1201 (1995).

3. Q.-Y.Tong, R.Scholz, U.Gosele, T.-H.Lee, L.-J.Huang, Y.-L.Chao, T.Y.Tan, *Appl. Phys. Lett.* **72**, 49 (1998).

4. A. Agarwal, T.E. Haynes, V.C. Venezia, D.J. Eaglesham, M.K. Weldon, Y.J. Chabal, O.W. Holland, in *Proceedings of 1997 IEEE Int. SOI Conference,* p.44, IEEE, Piscataway, NJ (1997).

5. T. Hoshbauer, M. Nastasi, J.W. Mayer, *Appl. Phys. Lett*, **75**, (2000).

6. J.A. Bardwell, L. le Brun, R.J. Evans, D.G. Curry, R. Abbott, *Review of Scientific Instruments,* **67**, 2346, (1996).

7. P. de Mierry, A. Etcheberry, M. Aucouturier, *Physica B: Condensed Matter*, **170**, 124, (1991).

8. S.J. Pearton, W.L. Hansen, E.E. Haller, J.M. Kahn, *Journ. of Appl. Phys.,* **55**, 1221, (1984).

9. G.S. Oehrlein, J.L. Lindstrom, J.W. Corbett, *Physics Letters*, **81A**, 246, (1981).

10. M.N.V. Raghavan, V. Venkataraman, *Semicond. Sci. and Technol.,* **13**, 1317, (1998).

11. S. Ashok, in *Proceedings of 1998 IEEE International Conference on Solid-State and Integrated Circuit Technology*, p.749, IEEE, Piscataway, NJ (1998).

12. F.A. Reboredo, M. Ferconi, S.T. Pantelides, *Physical Review Letters*, **82**, 4870, (1999).

13. S. M. Myers M. I. Baskes H. K. Birnbaum J. W. Corbett G. G. DeLeo S. K. Estreicher E. E. Haller P. Jena N. M. Johnson R. Kirchheim S. J. Pearton M. J. Stavola, *Reviews of Modern Physics,* **64**, 559, (1992).

14. G.F. Cerofolini, G. Calzonari, F. Corni, C. Nobili, G. Ottaviani, R. Tonini, *Materials Sci. And Eng. B,* **B71,** 196 (2000).

15. G.F. Cerofolini, F. Corni, S. Frabboni, C. Nobili, G. Ottaviani, R. Tonini, *Materials Sci. And Eng. Reports,* **R27***,* 1 (2000).

16. N. H. Nickel, G. B. Anderson, N. M. Johnson, J. Walker**,** *Phys. Rev. B, **62**, 8012(2000).*

17. A.Y. Usenko, "Process for lift-off a layer from a substrate" *US Patent Pending* 09/578896, 01/06/2000.

18. A.Y.Usenko, W.N. Carr, in *Proceedings of 2000 IEEE SOI Conference*, p.16, IEEE, Piscataway, NJ (2000).

Mat. Res. Soc. Symp. Proc. Vol. 681E © 2001 Materials Research Society

Electrical characterisation of UHV-bonded silicon interfaces

A. Reznicek, S. Senz, O. Breitenstein, R. Scholz and U. Gösele
Max-Planck-Institute of Microstructure Physics
Weinberg 2, D-06120 Halle, Germany

Abstract

Direct wafer bonding can be used to mechanically and electrically connect semiconductors. In our experiments two 100 mm diameter (100) Si wafers (n-doping: 10^{14} cm^{-3}) are first cleaned by standard chemical cleaning (RCA 1, 2). The surface is terminated by hydrogen after a HF dipping. The wafers are prebonded in air to protect the surface. After introduction into the ultra high vacuum (UHV) system the wafers are separated again. The hydrogen termination is released in a heating chamber. RHEED confirmed a surface reconstruction. The wafers are then cooled down to room temperature and bonded in UHV. The bonding energy is very close to the bulk bonding energy.

Measurements of whole n-n wafers showed a linear relationship of voltage and current at a low current density of 0.05 A/cm^2. The current flow is inhomogeneous, which is visible in IR-thermography images. Above 0.1 V the current density first saturates, but increases super-linearly for higher voltages. The electrical properties of a grain boundary can be modeled by a double Schottky barrier. The barrier height decreases with increasing applied voltage. C-V measurements show a strong dependence of capacitance on frequency, temperature and applied voltage.

The capacitance increases with higher temperature and lower frequency. The interface state density can be estimated from the low temperature and high frequency capacitance limit as $D_{it} = 1 \cdot 10^{11}$ cm^{-2}eV^{-1} assuming a constant density of states.

We can conclude that in order to avoid the undesirable effect of the potential barrier and trap states at the bonding interface a high doping near the interface is required for the application of wafer bonding to devices with a high current density across the bonded interface.

Introduction

Grain boundaries in silicon and other semiconductors have been investigated since many years. Most of the interest concentrates on materials for polycrystalline solar cells or voltage dependent resistors (varistors). The electrical properties of these devices are determined by the grain boundaries [1, 2]. The usual processing includes a high temperature step, e.g. solidification of a silicon melt to produce solar cells. Due to this high temperature process most of the impurities are concentrated at the grain boundaries. This results in a high density of electrically active states and a double sided depletion layer. In our experiments we produce an artificial grain boundary by bonding of two silicon single crystals. The most important distinction to other experiments is that we create this grain boundary at room temperature. The bonding is performed in UHV and allows direct formation of covalent bonds.

Experimental

For the experiments 100 mm CZ-grown silicon wafers were used. The p-doping is $1 \cdot 10^{15}$ cm^{-3} and the n-doping is $3 \cdot 10^{14}$ cm^3. On the backside the wafers are higher doped. A 1 µm thick epitaxial layer ($1 \cdot 10^{18}$ cm^{-3}) is grown to contact the silicon electrically. The wafers were first cleaned by standard chemical cleaning in RCA 1 and 2 solutions. After cleaning the wafers are rinsed with deionised water (18 MΩ cm). Afterwards the oxide layer is removed by dipping in hydrofluoric acid (2%), leaving a hydrophobic surface. Now the surface is terminated by hydrogen and the wafers are prebonded in a class 1 cleanroom to protect the surface. The wafer pair is transferred into an UHV system (10^{-10} mbar). Inside of this system the wafers are separated again and heated to remove the hydrogen from the surface. The hydrogen termination is released at 450°C for 5 minutes. After cooling down to room temperature the wafers are bonded without applying external pressure. For the electrical contacts an aluminum layer is evaporated on both epitaxial backsides. After the thermography measurements the wafer pair is cut in pieces and characterized in detail.

Results and Discussion

During heating the desorption of the hydrogen is monitored with a quadrupole mass spectrometer. The 2×1 reconstruction of the silicon surface is observed with RHEED. The wafers bond completely at room temperature. An IR-transmission image of a nn-wafer pair shows the bond quality (figure 1). Near the center a particle is enclosed and produces an unbonded area. The electrical measurement of the complete wafer pair is performed with a current flowing across the interface. Measurements of a whole n-n wafer pair show a linear relationship of voltage and current at low current density of 0.05 A/cm^2. The linear part of the I-V curve is non-ohmic. Lock-in Thermography [3] images show an inhomogeneous distribution of dissipated heat (figure 2). Most of the current flows on the rim of the bonded region.

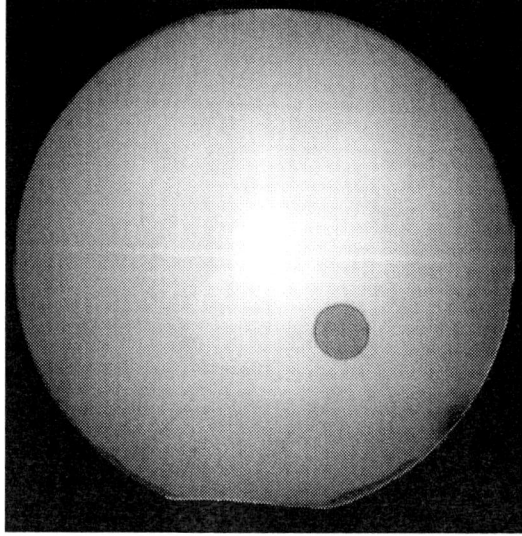

Figure 1: IR-transmission image of a bonded n-n wafer pair

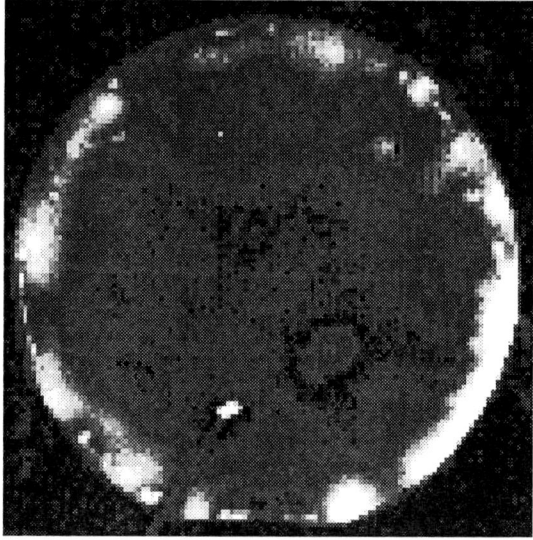

Figure 2: Thermography image of the n-n wafer pair (white → high current flow)

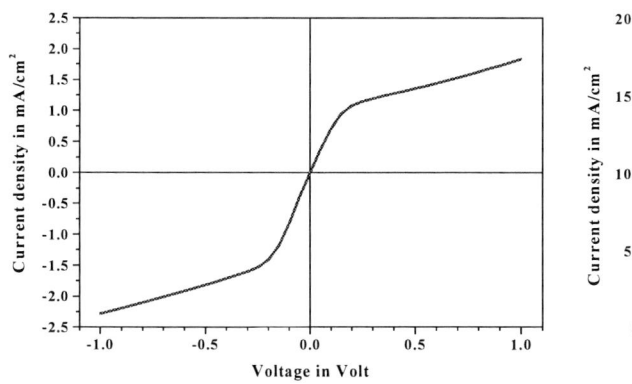
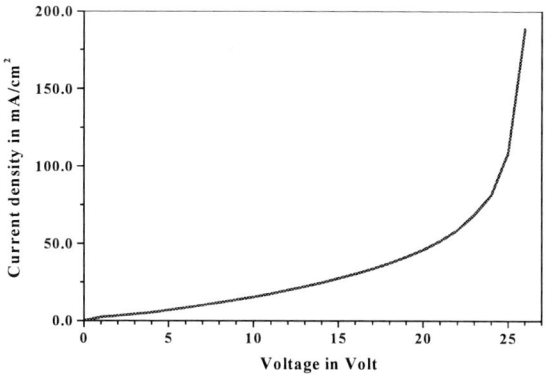

Figure 3 : nn-junction: low voltage I-V curve   Figure 4: I-V curve up to breakdown (n-n)

The bonding quality of the rim region is deteriorated by small particles introduced during splitting of the prebonded wafer pair and handling procedures inside the UHV chamber. Therefore pieces from the center region were prepared. The I-V curve shows a saturation above 0.1 V and increases super linearly for higher voltages (figures 3 and 4). At voltages lower than approximately 5 V the current flow can be described by thermionic emission over a barrier, which is formed due to states at the interface [4]:

$$J_{th} = A^* T^2 \cdot e^{-\frac{\zeta + \Phi_B}{kT}} \cdot \left(1 - e^{-eV/kT}\right) \tag{1}$$

J_{th} is the emission current density, A^* the effective Richardson constant, T the temperature, ζ is the energy separation between the Fermi level and the conduction band in the neutral crystallites and Φ_B is the barrier height (see figure 5).

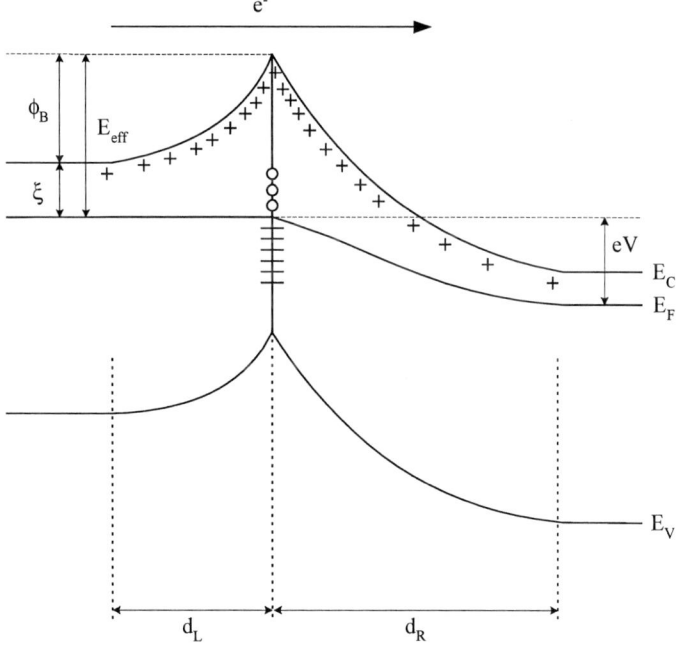

Figure 5: Energy-band diagram for an nn-junction under applied voltage

59

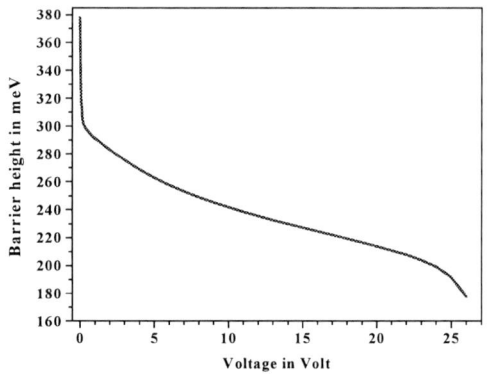

Figure 6: Voltage dependence of barrier height for an nn-junction

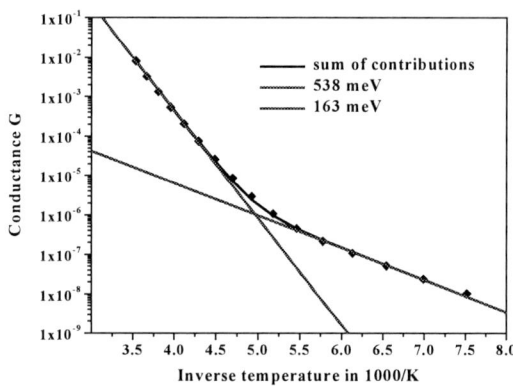

Figure 7: Temperature dependence of conductance (p-p)

d_L and d_R describe the widths of the depletion layer on both sides. The effective barrier height E_{eff} is defined as:

$$E_{eff} = \Phi_B + \xi \tag{2}$$

The voltage dependence of the barrier height is calculated from the I-V data and equation 1 (figure 6). The barrier height decreases with increasing applied voltage. At higher voltages the electrical field in the space charge region is higher than the threshold of electron hole pair generation. This effects is responsible for an additional current density of minority carriers which further decreases the barrier height [5]. The barrier heights at room temperature and zero voltage are determined from the measured current density based on equation (1) for the nn-junction as 380 meV and from analogous measurements for the pp-junction as 300 meV. From temperature dependent low voltage conductance measurements (eV<< kT) we calculate two different effective barrier heights for the pp-junction (see figure 7). Using equation (2) the barrier height Φ_B at room temperature can be calculated as 290 meV.

C-V measurements show a strong dependence of capacitance on frequency, temperature and applied voltage as shown in figure 8. At low temperatures the states at the interfaces do not respond to the applied AC voltage and show the usual $1/C^2$-behavior of Schottky-diodes.

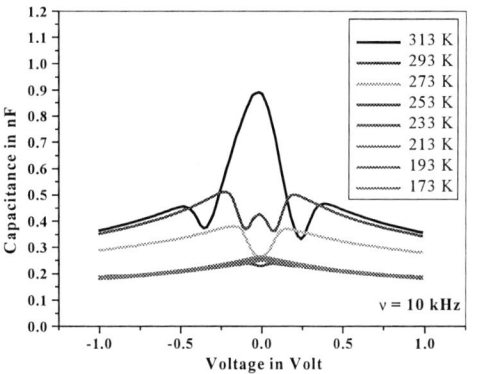

Figure 8: Capacitance vs. bias voltage (n-n)

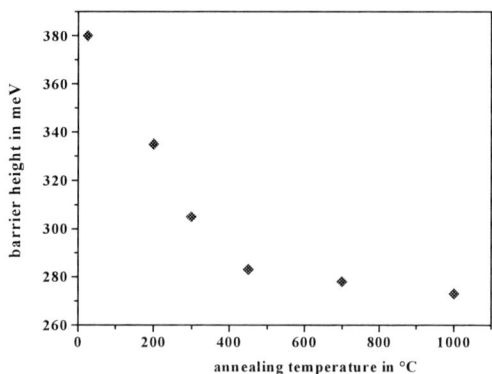

Figure 9: barrier height of heated nn-samples

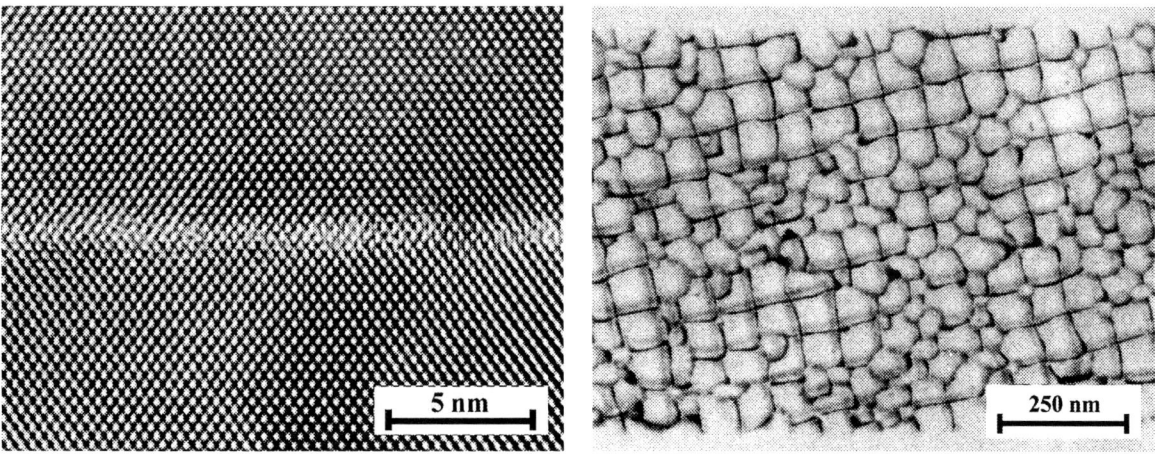

Figure10: HR-TEM cross-section of a bonded nn-junction

Figure 11: TEM Plan view on a bonded and annealed nn-junction (1000°C for 2 h)

The features developing at higher temperatures originate from the filling and emptying of the interface states. The bias-voltage dependence of the capacity is related to the distribution function of the interface states. This effect is described by the "Trap Transistor Model" [6]. During the C-V measurement a small AC voltage is applied and modulates the barrier height. The resulting current density shows a strong modulation (equation 1). The phase of the resulting AC-current density is shifted, the 90° component is displayed as capacitance. The interface state density can be estimated from the low temperature and high frequency capacitance limit as $D_{it} = 1 \cdot 10^{11}$ cm^{-2}eV^{-1} assuming a constant density of states.

The electrical application of bonded structures for high power switches or lasers requires high current densities at low voltages. The current density can be increased by annealing the UHV bonded interface (figure 9) or by higher doping at the bonded interface. The room temperature bonded interface is not in thermodynamical equilibrium and contains a high density of near atomic scale defects (figure 10). During heating most of the defects are removed and at temperatures above 800°C a dislocation network is formed (figure 11).

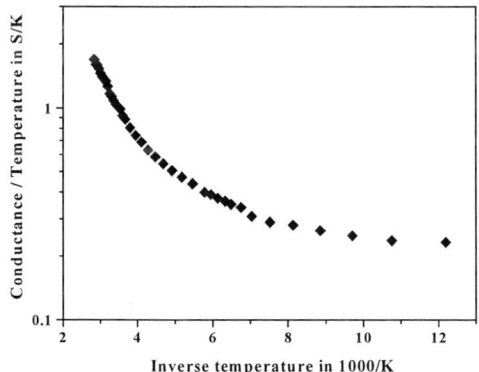

Figure 12: Temperature dependence of conductance for a highly doped nn-junction

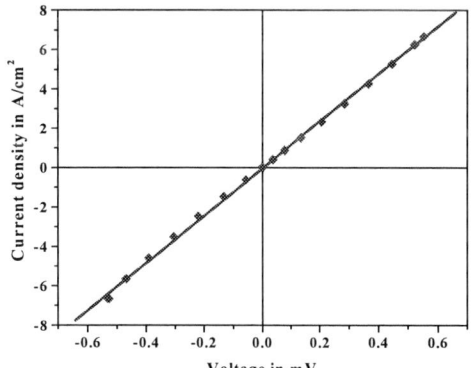

Figur 13: I-V characteristics for the highly doped nn-juntion

If the interface is highly doped ($1 \cdot 10^{19}$ cm^{-3}) the current density increases strongly. The barrier height and the width of space charge region decrease. The barrier height can be estimated from theoretical calculations as 60 meV. Thermally assisted tunneling across the barrier contributes to the current density. Figure 12 shows the non-Arrhenius temperature dependence of conductance expected for a sample with a high tunneling contribution. The voltage drop at the interface is 0.5 mV for a current density of 6 A/cm^2 (figure 13).

Conclusions

In order to avoid the undesirable effects of the potential barrier and interface states at the bonding interface a high doping near the interface is required for application of wafer bonding to devices with a high current density across the bonded interface.

References

[1]   G.E. Pike, Phys. Rev. B **30**, 795 (1984)

[2]   G. Blatter and F. Greuter, Phys. Rev. B **33**, 3952 (1986)

[3]   O. Breitenstein, M. Langenkamp, F. Altmann, D. Katzer, A. Lindner, and H. Eggers, Rev. Sci. Inst. **71**, 4155 (2000)

[4]   S.M. Sze, in Physics of Semiconductor Devices, (Wiley Intersciences New York, 1981), p. 255 – 258

[5]   G. Blatter and F. Greuter, Phys. Rev. B **34**, 8555 (1986)

[6]   J. Werner, K. Ploog and H.J. Queisser, Phys. Rev. Let. **57**, 1080 (1986)

Mat. Res. Soc. Symp. Proc. Vol. 681E © 2001 Materials Research Society

Ge Layer Transfer To Si For Photovoltaic Applications

James M. Zahler, Chang-Geun Ahn, Shahrooz Zaghi, and Harry A. Atwater
Thomas J. Watson Laboratory of Applied Physics, California Institute of Technology,
Pasadena, CA 91125, U.S.A.

Charles Chu and Peter Iles
Tecstar, Inc.,
City of Industry, CA 91745-1002, U.S.A.

ABSTRACT

We have successfully used hydrophobic direct wafer bonding along with hydrogen-induced layer splitting of germanium to transfer 700 nm thick, single-crystal germanium (100) films to silicon (100) substrates without using a metallic bonding layer. The metal-free nature of the bond makes the bonded wafers suitable for subsequent epitaxial growth of layered solar cells at high temperatures without concern about metal contamination of the device active region. Contact mode atomic force microscopy images of the transferred germanium surface generated by the formation of micro-bubbles and micro-cracks along the hydrogen-induced layer-splitting interface reveals minimum rms surface roughness of between 10 nm and 23 nm. Electrical measurements indicated ohmic I-V characteristics for germanium layers bonded to silicon substrates with ~400 Ω cm^{-2} resistance at the interface. Triple-junction solar cell structures grown on these Ge/Si heterostructure templates by metal-organic chemical vapor deposition show comparable photoluminescence intensity and minority carrier lifetime to a control structure grown on bulk Ge. The use of a molecular beam epitaxy Ge buffer layer to smooth the cleaved surface of the Ge heterostructure has been shown to smooth the rms surface roughness from ~11 nm to as low as 1.5 nm with a mesa-like morphology that has a top surface roughness of under 1.0 nm giving a promising surface for improved solar cell growth on solar cell structures.

INTRODUCTION

Compound III-V semiconductor layered structures grown on bulk germanium substrates have been used to create high efficiency triple-junction solar cells with efficiencies greater than 30% [1, 2]. However, these are prohibitively expensive for all but space applications. The Ge substrate constitutes a large portion of this cost. Ge/Si heterostructures formed by wafer bonding and layer transfer of a thin crystalline Ge layer by hydrogen-induced exfoliation are being considered as a way to reduce the product cost while maintaining solar cell device performance. By transferring thin, single-crystal layers of Ge to a less expensive Si substrate and reclaiming the donor wafer through a polish process, a single 300 μm thick Ge wafer could serve as a source for transfer of in excess of 100 thin Ge layers.

This application requires the bond at the interface of the Ge/Si heterostructures to be covalent to ensure good thermal contact, mechanical strength, and to enable the formation of ohmic contact between the Si substrate and Ge layers. To accomplish this hydrophobic wafer bonding will be used, because the H surface-terminating species that facilitate van der Waals bonding have been shown to be evolvable from the surface in Si/Si bonding systems at temperatures above 600 °C [3, 4]. We believe that this phenomenon will be extendable to hydrophobically bound Ge/Si layer transferred systems.

EXPERIMENTAL DETAILS

(100) Ge wafers were implanted with H+ at 80 keV and 1.0×10^{17} cm^{-2} [5]. These wafers are rendered strongly hydrophobic in the process. Ge passivation then consists of removal of adsorbed organic contamination and maintaining the hydrophobic nature of the H-implanted Ge. The Ge is cleaned by acetone and methanol to remove organic contaminants. This is followed by a one minute DI rinse and a subsequent 10-second dip in 5% HF and surface drying to remove any remaining oxide. The wafers are then blown dry.

(100) Si wafers were passivated by the same wet process sequence described above and a subsequent rinse followed by a DI rinse and a 30 second 80 °C 1:1:3 H_2O_2:NH_4OH:H_2O (RCA1) clean process followed by a DI rinse and a brief HF dip to remove the grown oxide. The RCA1 clean is included to further reduce the organic surface contamination and remove particles. Following surface passivation both Si and Ge have and rms roughness well under 0.5 nm as measured on a 5 μm by 5 μm atomic force microscopy scan.

After passivation the wafers are brought into contact at room temperature and bonding is initiated by a 3500-psi pressure applied over a 0.25-inch diameter region at the center of the wafer. The contact region is then propagated outward using subsequent pressures of 890-psi applied over a 0.5-inch diameter region and 220-psi applied over a 1.0-inch diameter region.

A thermal process to 175 °C with an applied pressure of 135 psi in a modified Parr Instruments pressure cell is used to strengthen the bonding. Layer splitting is achieved by the formation of hydrogen-containing platelets that initiate the propagation of micro-cracks parallel to the Ge surface upon annealing to greater than 350 °C with no external pressure [5].

Metal-organic chemical vapor deposition growth of triple-junction solar cells on Ge/Si heterostructures was performed in collaboration with Tecstar. The peak temperature of the cycle was in excess of 750 °C and the structure consists of a GaAs buffer layer followed by two active base regions – one in GaAs and one in InGaP – separated by proprietary tunnel-junction structures. A heavily doped top contact layer of GaAs was used to probe for photoluminescence intensity and minority carrier lifetime in both a control sample grown on bulk Ge and structures grown on Ge/Si heterostructures.

DISCUSSION

Wafer bonding and layer transfer has been achieved, but initial efforts were frustrated by the formation of blisters at the bonded interface. These bubbles likely caused by contamination of the interface at the time of bonding, either by adsorbed water that evolves with temperature or organic contamination that decomposes with temperature to evolve hydrogen. [6]

These blisters have been minimized by two separate techniques. First, by growth of a 40 Å amorphous Si layer on the H-implanted Ge substrate, hydrophilic wafer bonding can be achieved by using the same hydrophilic surface passivation on both substrates. This bonding energy has been well studied for the formation of silicon on insulator by bonding hydrophilic Si surfaces to hydrophilic thick oxides and is typically ~100 mJ/cm^2 for Si/Si systems [4]. The Ge/Si heterostructures formed by Si/a-Si hydrophilic bonding show a strong reduction in the total number of interfacial bubbles. This is thought to be due to the increased hydrophilic bond strength and the improved organic removal made possible by the RCA1 clean. Additionally, the SiO_2 layer is thought to accommodate desorbed gas that would otherwise be trapped at the interface.

A second approach to eliminate bubbles at the interface is the use of a 250 °C pre-bonding anneal in N_2 prior to bringing the surfaces together. This pre-bonding anneal is thought to desorb water and degrade organic contamination and desorb it from the surface leaving a surface that is more perfectly H-terminated. This has two competing effects; transferred layers have a reduction in defects, but the room-temperature bond between the wafers is more difficult to initiate.

Ohmic Electric Contact

The electrical properties of the interface were measured by evaporating aluminum contacts on a Ge/Si heterostructure prepared by a pre-bonding anneal in N_2 as previously described and a layer split anneal at 350 °C. The Ge substrate was doped with $5x10^{17}$ cm^{-3} Gallium and the Si substrate was doped with 10^{18} to 10^{20} cm^{-3} Boron in effort to minimize the potential barrier and depletion width of any junction that might be formed at the heterojunction interface.

The first scan from –10 V to 10 V showed a dielectric breakdown followed by ohmic I-V characteristics in subsequent scans (Fig. 1). These measurements indicate an interfacial resistance of 35 to 40 Ω over a total interfacial area of ~0.1 cm^2 for an areal interfacial resistance of ~400 Ω cm^{-2}. The contact and substrate resistances were determined to be negligible for overall structure resistance.

This structure resistance is extremely high for integration into a high efficiency solar cell; however, the measured sample was annealed to a much lower temperature than the anticipated covalent bond formation temperature of 600 °C or greater. Thus, there may be a removable dielectric layer of bound hydrogen and sub-monolayer amounts of other contaminants at the interface. Future tests made by varying the peak anneal temperature along with TEM analysis of the interface will study the resistance dependence on annealing conditions and the interfacial results of those annealing conditions.

Triple Junction Solar Cell Growth

Triple-junction solar cell structures were grown by metal-organic chemical vapor deposition on Ge/Si heterostructures made by hydrophobic wafer bonding. Two Ge/Si heterostructures were used as templates for growth, labeled Sample 1 and Sample 2, while a bulk Ge control solar cell structure was also grown in the same process. The surface roughness of these structures was measured by contact mode atomic force microscopy. The results of this analysis

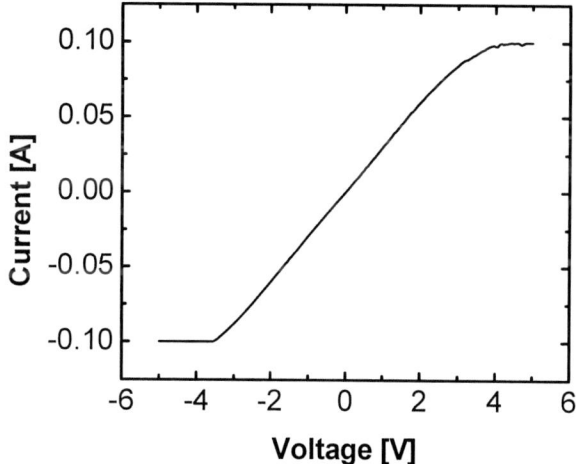

Figure 1. Current-Voltage curve for a P+ Ge / P+ Si heterostructure annealed to 350 °C.

Table 1. MOCVD triple-junction solar cell structure roughness measurements.

Sample Name	Pre-MOCVD Ge Roughness (Å)	Post-MOCVD GaAs Roughness (Å)
Bulk Ge	<5	147
Sample 1	236	897
Sample 2	225	204

are given in Table 1. The two bonded heterostructures exhibited dramatically different roughness of the grown GaAs contact layer. The reason for the disparity in the surface roughness between the two Ge/Si heterostructures is not presently understood.

Cross-sectional scanning electron microscope images were made of Sample 1 and the bulk Ge control structure (Fig. 2). These images show the layer structure of the triple-junction solar cell and the morphology of the interfaces of the various layers and top surface. The control image shows smooth layer transitions to within the resolution of the microscope (~100 nm). The Ge/Si heterostructure (Sample 1) template is shown to have a rough interface between the layers of the cell structure with maximum surface undulations of ~1 μm at the GaAs interfaces.

Photoluminescence studies of the top GaAs contact layer using a 458 nm Ar laser pump probe indicate comparable GaAs band edge emission at 880 nm between the bulk Ge control and Sample 2, the smoother grown structure (Fig. 3). The limited on the roughness of the GaAs surface and photoluminescence measurements indicates an inverse relationship between surface roughness and GaAs signal intensity as an indirect indication of defect density in the grown III-V structure.

Time resolved photoluminescence measurements of the GaAs cap layer performed at NREL indicate short but comparable decay time constants of 0.23 ns for the bulk Ge sample and 0.20 ns for Sample 2 indicating a comparable minority carrier lifetimes assuming consistent surface recombination velocities. The presence of an exposed high recombination site density GaAs top surface limits the lifetime of minority carriers in the top surface of this GaAs cap layer. Additionally, the GaAs cap layer of the solar cell structure is deposited for the purpose of making ohmic metal contacts to the solar cell structure and for that reason is heavily doped thereby increasing the density of recombination sites for electron-hole pairs. Future tests are

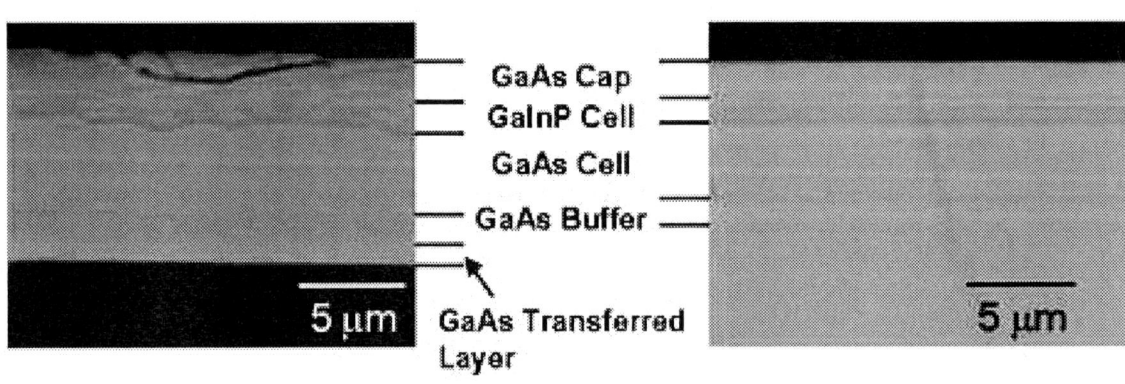

Figure 2. Cross-sectional SEM image of MOCVD triple-junction solar cell structure as grown on a Ge / Si heterostructure template (left) and a bulk Ge substrate (right).

planned to obtain PL and minority carrier lifetimes by probing the confined GaAs base region in the solar cell structures that lies between two layers of InGaP that should serve to confine electron-hole pairs to the GaAs layer while eliminating the surface state recombination sites.

Ge Surface Smoothing with Ge MBE Buffer Layer

The triple-junction solar cell optical performance results indicate that without any surface preparation following the H-induced cleavage of the Ge layer, high quality III-V photovoltaic

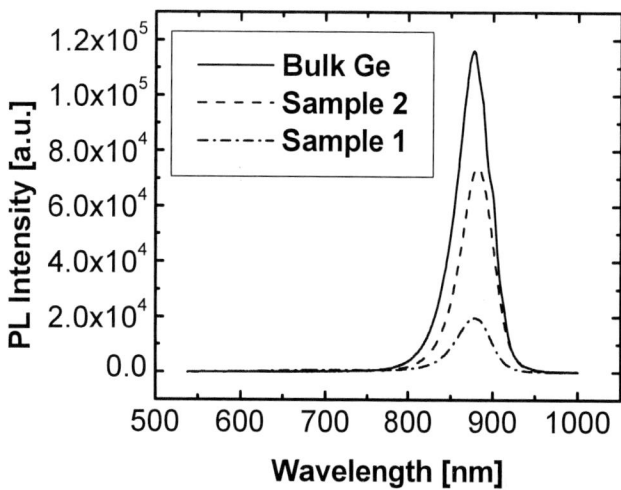

Figure 3. GaAs band edge emission photoluminescence of MOCVD grown triple-junction solar cells on Ge / Si heterostructures and on a bulk Ge control wafer.

materials can be grown with good PL and minority carrier lifetime properties relative to a cell grown on a bulk Ge substrate. However, to further improve the optical and electrical properties it is desirable to be able to smooth the morphology of the exfoliated surface. To smooth the exfoliated Ge surface a 2500 Å thick Ge buffer layer was grown on the surface of the Ge/Si heterostructure. This reduced the roughness of the transferred layer from about 110 Å to 22.4 Å. Additionally, the morphology of the surface drastically changed to a mesa-like form a large relatively smooth layer with less than 10 Å surface roughness. Deep pits decorate this mesa-like surface and are thought to be threading dislocations (Fig. 4). The RHEED pattern following the growth also indicated a smooth, reconstructed Ge surface (Fig. 5).

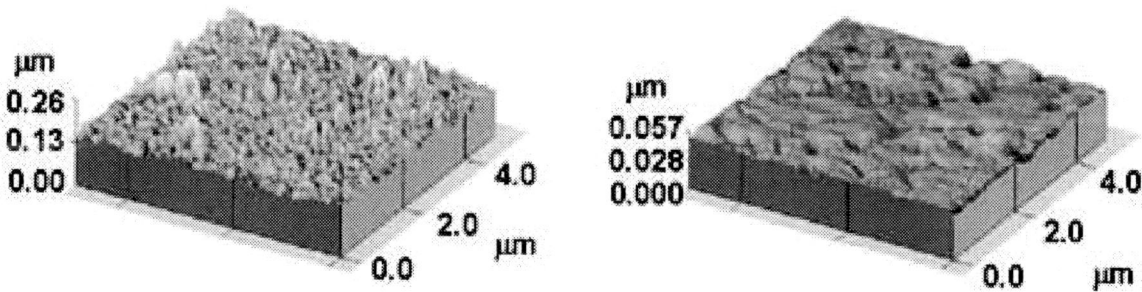

Figure 4. Exfoliated Ge surface prior to MBE Ge buffer layer growth (left) with 110 Å rms roughness, is smoothed to just 22.6 Å rms roughness with a mesa geometry (right).

CONCLUSIONS

This work shows the ability to make ~1 cm^2 area Ge/Si heterostructures by hydrophobic wafer bonding followed by H-induced layer splitting. These heterostructures have been shown to be both ohmic and acceptable templates for MOCVD growth of triple-junction solar cell structures with photoluminescence and lifetime properties comparable to those found in solar cell structures grown on bulk Ge. The technique of Ge buffer layer smoothing shows strong promise as a means of further improving the optical qualities of the MOCVD grown III-V compounds in the solar cell structure.

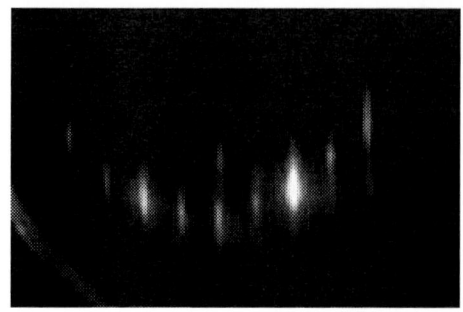

Figure 5. Post-growth RHEED image of the surface showing Bragg rods and a reconstructed surface indicating a smooth top plateau.

ACKNOWLEDGEMENTS

The authors would like to thank Aditi Risbud for her help in imaging of cross-sectional SEM samples and Peter Kik in his help in setting up the photoluminescence measurement equipment. Additionally, Dick Ahrenkiel at NREL is thanked for is help in obtaining minority carrier lifetime measurements.

REFERENCES

1. S. M. Sze *Physics of Semiconductor Devices* (Wiley and Sons, New York, 1981) pp. 790-799.
2. M. A. Green "Solar Cells," *Modern Semiconductor Device Physics,* edited by S. M. Sze (Wiley and Sons, New York, 1998) pp. 503-512.
3. M. K. Weldon, Y. J. Chabal, D. R. Hamann, S. B. Christman, E. E. Chaban, and L. C. Feldman, *J. Vac. Sci. Technol. B* **14** (4) 3095 (1996).
4. Q.-Y. Tong, E. Schmidt, U. Gosele, and M. Reiche, *Appl. Phys. Lett.* **64** (5) 625 (1994).
5. Q.-Y. Tong, K. Gutjahr, S. Hopfe, T.-H. Lee, and U. Gosele, *Appl. Phys. Lett.* **70** (11) 1390 (1997).
6. S. Mack, H. Bauman, U. Gosele, H. Werner, and R. Schlogl, *J. Electrochem. Soc.,* **144**, 1106 (1997).

Mat. Res. Soc. Symp. Proc. Vol. 681E © 2001 Materials Research Society

Anodic Bonding at Room Temperature

Volker Baier[1], Andreas Gebhardt[2] and Stefan Barth[3]
[1]Institute for Physical High Technology, Micro Systems Division, Winzerlaer Straße 10,
D-07745 Jena, Germany, e-mail: baier@ipht-jena.de
[2]VITRON Spezialwerkstoffe GmbH, Otto-Schott-Straße 13, D-07745 Jena, Germany,
e-mail: kontakt@vitron.de
[3]Hermsdorfer Institut für Technische Keramik e. V., Magnetwerkstoffe, Marie-Curie-Straße 17,
D-07629 Hermsdorf, Germany

ABSTRACT

Special phosphate glasses (niobium-phosphate glasses) are used for anodic bonding experiments. They are optimized particularly with regard to elevated alkali ion conductivity at room temperature due to high contents of Li or Na ions. The compounds anodically bonded at room temperature show the strength of the breakage of the glass. Only a few seconds up to a few minutes at voltages of maximum 500 V are necessary for bonding. The glasses have a thermal expansion coefficient, which is not fitted to that of silicon. Therefore, stability of the bonded compounds is achieved only at temperature jumps up to 100 K in the case of bulk samples. For higher temperature jumps the probability of breakage increases continuously. It is possible to get thin films of these glasses with ac and dc magnetron sputter techniques for example on silicon. These thin layers are anodically bondable and the compounds with silicon show a higher temperature stability (at least 233 to 473 K).

INTRODUCTION

Commonly used glasses for anodic bonding like Pyrex bond glass (# 7740) are bonded typically at 623 - 673 K. With Li-aluminosilicate-β-quartz glass-ceramic (HOYA # PS-100)[1] and Li-alumino-boro-silicate glass[2] bond temperatures of 473 K are reached.

But anodic bonding at lower temperatures gets more and more important as a method for hermetic encapsulation of many components, because this process is one of the last of the many manufacturing steps and, therefore, high temperatures are often not allowed.

For the anodic bonding process it is necessary to have a high ion conductivity in the bond glass and the thermal expansion coefficients of the bond partners should be well fitted to each other (for example silicon to the bond glass) because of the temperature range of operation. On the other hand, the condition of thermal fitting of the expansion coefficients is of lower importance, if it is possible to bond at ambient temperature. Therefore, the search for glasses with high ion conductivity at room temperature is of main interest.

EXPERIMENT

Preparation of glasses

Two glasses are prepared in the system $R_2O-Nb_2O_5-P_2O_5$ with R = Li or Na. The composition is optimized for high alkaline-ion conductivity at room temperature[3].

The reagents Na_2CO_3, $NH_4H_2PO_4$ and Nb_2O_5 are mixed and then the homogeneous powder is slowly moistened in a beaker with some water. Now the formation of CO_2 begins. After the end of the degassing the slurry is heated at 353 K for about 18 h.

The glasses are obtained by melting of the raw materials at 1273 K in a Pt crucible for about 2 h. Finally, the melts are cast into a fine grained dense graphite mould followed by cooling from 723 K to room temperature at about 5 Kmin^{-1}.

Table I. Materials for low temperature anodic bonding

	Temperature (K)	Electrical conductivity (Ω^{-1}cm^{-1})	Thermal expansion coefficient (10^{-6} K^{-1})
Silicon (for comparison)	298		3.2
Pyrex glass (Corning # 7740)[1]	523	7.7 x 10^{-9}	3.3
HOYA bonding glass (# SD-2)[1]	473	2.6 x 10^{-10}	3.2
Li-aluminosilicate-β-quartz glass-ceramic (HOYA # PS-100)[1]	473	2.7 x 10^{-8}	3.1
Li-alumino-boro-silicate glass[2]	473		3.4
Na-niobate-phosphate glass[4]	298	4 x 10^{-8}	17.6
Li-niobate-phosphate glass[4]	298	9 x 10^{-8}	12.2

Thin sputtered glass films

Low temperature niobate-phospate bond glass is sputtered on 4 inch silicon wafers by HF magnetron sputter technique (Alcatel PUMA 500). For an 1 μm glass layer 300 W are necessary for 2 hours at 5 x 10^{-3} mbar and 40 sccm Ar flow. Higher power rates lead to partially destroying the glass target and generation of particles.

Anodic bonding conditions

Bonding experiments are carried out in an anodic bonding chamber AB1 of Electronic Visions enabling bonding under vacuum conditions. Bulk bond glass material of 1.5 mm thickness is bonded on silicon at 300 K and 300 V during 600 s. The electric charge flow is about 1 Ccm^{-2}. For bonding of thin films (thickness 1 μm) on silicon at 300 K, 250 V with a duration of 300 s is necessary leading to an electrical charge flow of about 2 Ccm^{-2}.

Resistance to thermal shock of the bond compounds

In a climate chamber alternating temperature variations are carried out beginning at room temperature + 10 K → - 20 K → + 30 K → - 40 K → + 50 K → + 60 K → + 70 K • • • .

RESULTS

Niobate-phosphate glasses with high ion conductivity at room temperature are suitable for room temperature anodic bonding. Bond compounds of silicon to low temperature phosphate bond glass (10 x 10 mm^2, thickness 1.5 mm) withstand temperature cycles of ± 50 K. The glass breaks at higher temperature changes.

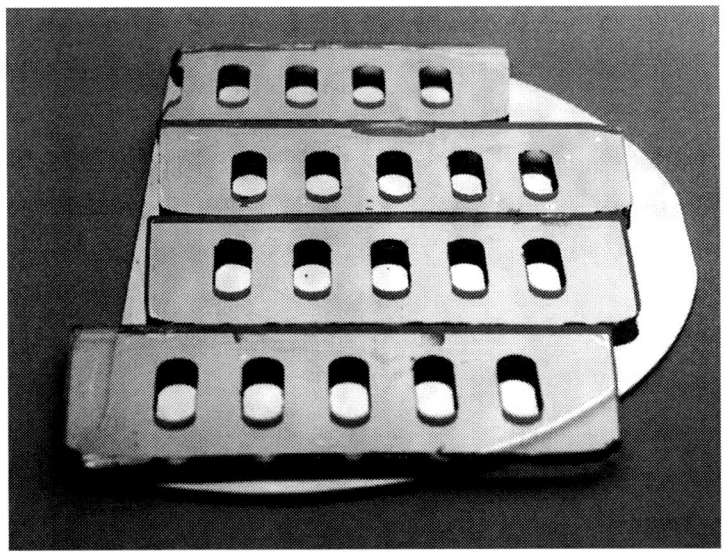

Figure 1. Anodic bond samples - glass on silicon

Figure 2. Anodic bond sample silicon to silicon sputter-deposited with 1 μm low temperature niobate-phospate bond glass 20 x 20 mm^2

Niobate-phosphate glasses can be sputtered by dc and ac magnetron sputtering technique. Bond probes silicon to silicon sputter-deposited with 1 μm low temperature phosphate bond glass (20 x 20 mm^2) are stable at least from 233 to 473 K even at high temperature change rates (directly put in and removed out of the oven).

An example of anodic bonding of niobate-phosphate glass on silicon is shown in figure 1. Figure 2 shows an anodic bond sample of silicon to silicon sputter-deposited with niobate-phosphate glass film (thickness 1 μm).

CONCLUSIONS

Anodic bonding at room temperature with new niobate-phosphate bond glasses as bulk or sputtered thin film has been performed. This anodic bonding technique should be widespread applicable to the hermetical sealing of thermally sensible samples like low boiling liquids (e.g. water), biological samples or, in general, components that do not withstand higher temperatures.

ACKNOWLEDGEMENTS

The investigations are supported by the Thüringer Ministerium für Wissenschaft, Forschung und Kunst (collaboratory project ANNI grant number B609-98025). Many thanks to S. Schundau (AMT Jena) for sputtering technique and M. Arnz, A. Franke, F. Jahn, I. Menzel, J. Müller, H. Porwol, H. Schellhorn, R. Stöpel and further colleagues from the IPHT Jena for technical support.

REFERENCES

1 S. Shoji, H. Kikuchi, H. Torigoe, Sensors and Actuators **A 64**, 95-100, (1998)
2 V. Baier, D. Hülsenberg, K. Schmidt, B. Straube, DE Patent 195 45 422 (6 Dezember 1995)
3 S. Barth, A. Feltz, Solid State Ionics **34,** 41-45 (1989)
4 V. Baier, A. Gebhardt, S. Barth, DE Patent pending, submitted on 27.10.2000

Mat. Res. Soc. Symp. Proc. Vol. 681E © 2001 Materials Research Society

Si/GaAs heterostructures fabricated by direct wafer bonding

Viorel Dragoi, Marin Alexe, Manfred Reiche, Ionut Radu,
Erich Thallner[1], Christian Schaefer[1] and Paul Lindner[1]
Max Planck Institute of Microstructure Physics,
Weinberg 2, D-06120 Halle (Saale), Germany
[1]EV Group, St. Florian, A-4780 Schärding, Austria

ABSTRACT

Si/GaAs heterostructures were obtained by a low temperature direct wafer bonding (DWB) method which uses spin-on glass (SOG) intermediate layers. The use of intermediate SOG layers allows the fabrication of Si/GaAs heterostructures at processing temperatures lower than 200°C. The achieved bonding energy permits thinning down to a few microns of Si and GaAs wafers, respectively, using grinding procedures followed by chemical mechanical polishing (CMP). After thinning, the heterostructures sustained annealing temperatures of 450°C without damaging of the bonded interface. The above bonding procedure was successfully applied for bonding GaAs wafers to Si wafers with structured surfaces. A technology was developed based on this bonding method for producing universal GaAs-on-Si or Si-on-GaAs substrates.

INTRODUCTION

Monolithic integration of compound semiconductors into silicon technology would result in new applications in optoelectronics, microwave electronics, and high temperature electronics. A specific interest is focused on the fabrication of Si/III-V compound semiconductor heterostructures. The combination of high performance III-V compound semiconductor optoelectronic devices with the charge handling functionality of modern silicon circuitry would enable the fabrication of monolithically integrated optical interconnects which will increase considerably the speed of data processing and transmission [1]. Silicon is also an ideal supporting material for GaAs and other III-V compound semiconductors due to its superior mechanical strength, low weight, and high thermal conductivity [2].

Classical thin film deposition techniques like low temperature epitaxy, metalorganic chemical vapour deposition or molecular beam epitaxy were used for the fabrication of GaAs layers on Si substrates. Compared to bulk GaAs, GaAs thin films on Si substrate suffer from two major problems: i) the presence of high dislocation densities due to the 4.1 % lattice mismatch between Si and GaAs (typical $10^6 \div 10^8$ cm^{-2}, instead of 10^3 cm^{-2}, which is desired for device fabrication), and ii) the biaxial tensile stress generated in the plane of GaAs film during cooling from the deposition temperature due to the thermal mismatch (thermal expansion coefficient - TEC - of GaAs is almost double than the TEC for Si) [3].

Direct wafer bonding (DWB) can be a valuable solution for solving the lattice mismatch problem as long as this technique impose conditions only to the flatness, microroughness and the cleanliness of the surfaces and is not depending on the crystalline properties of the two materials. Thermal mismatch remains also an issue for DWB but in the last years low temperature bonding techniques were developed (vacuum bonding [4], plasma activation methods [5, 6], bonding with intermediate layers [7, 8]) which can be applied to diminish the stress thermally generated at the interface during annealing.

This paper presents a novel low temperature DWB method which has been successfully applied to fabricate Si/GaAs heterostructures.

EXPERIMENTAL

Three bonding procedures were evaluated: i) hydrophilic bonding of GaAs and Si followed by annealing up to 300°C, ii) O_2 plasma activation of surfaces before bonding followed by annealing up to 300°C, and iii) bonding *via* a spin-on glass (SOG) intermediate layer.

(100) oriented semiinsulating GaAs wafers, 100 mm diameter, and (100) oriented p-type Si wafers, 100 mm diameter were used for experiments. The Si wafers were cleaned before bonding using the standard chemical cleaning procedure with RCA 1 ($NH_4OH:H_2O_2:H_2O = 1:1:5$) and RCA 2 ($HCl:H_2O_2:H_2O = 1:1:5$) solutions. The GaAs wafers were first rinsed with deionized water for particle removal. Then the wafers were cleaned in a $NH_4OH:H_2O = 1:20$ solution for the hydrophilic bonding and bonding *via* SOG layers. For oxygen plasma activation, the GaAs wafers were only rinsed in deionized water. The bonding procedures were performed in a class 10 cleanroom.

For the plasma activation of the surfaces the two wafers were introduced into an O_2 plasma reactor for 5 minutes (oxygen pressure – 1.5 torr, radio frequency power – 600 W) and then bonded using the procedure described above.

Finally, the bonding *via SOG* was applied. SOG layers were deposited by a spin-on technique onto the Si wafers. The thickness of the SOG layer was measured using a profiler after a step was etched into the layer. The deposited wafers were bonded with GaAs wafers and then annealed at low temperatures.

The bonded interface was investigated using infrared (IR) transmission, scanning acoustic microscopy (SAM) and transmission electron microscopy (TEM). The surface energy was measured with the crack opening method and tensile testing. The tensile test was performed on small samples cut from the bonded wafer pairs (6 x 6 mm^2). IR spectroscopy was used to study the chemistry of the glass intermediate layer. The fracture surfaces resulted after the samples brake during the tensile test were investigated with atomic force microscopy (AFM) and scanning electron microscopy (SEM).

RESULTS AND DISCUSSION

First, GaAs wafers were hydrophilic bonded with Si wafers. The surface energy at RT was about 20 mJ/m^2 and increased to about 80 mJ/m^2 after annealing at 300°C in a time range from 10 to 30 hours. The interactions mediating the adhesion at RT are weak and allow the wafers to separate during heating. The process is reversible and the wafers bond again during cooling stage. An oxygen plasma activation of the surfaces increases the surface energy at RT to about 80 mJ/m^2. By annealing the bonded wafer pairs in nitrogen at temperatures up to 350°C, debonding occurs due to the high stress developed at the interface and sometimes the GaAs wafers shatter. The use of an intermediate SOG layer was recently proposed as a low temperature method for GaAs/Si bonding [9].

A 350 nm thick layer was deposited onto Si wafers by spinning of a commercially available silicate SOG precursor. The resulted films were baked in air at temperatures up to 180°C. The SOG coated wafers were RT bonded with GaAs wafers using the standard bonding procedure. The surface energy at RT was about 0.4 J/m^2, almost four times higher than the surface energy in case of Si/Si hydrophilic bonding. This very high surface energy can be a result of the chemistry at the SOG-GaAs interface, which is very different from the usual silicon-silicon bonding. Infrared spectroscopy measurements presented in figure 1 revealed the existence of CH$_3$ radicals in the SOG films baked at temperatures below 200°C.

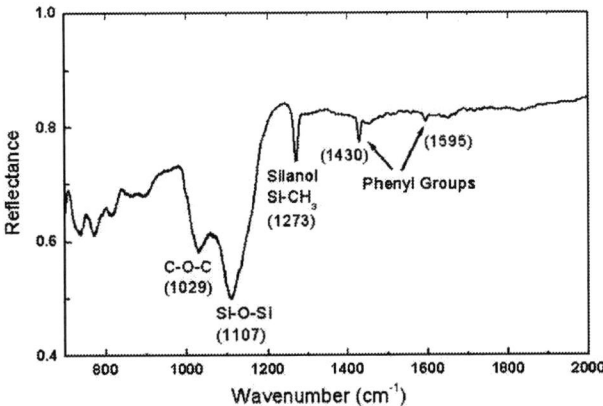

Figure 1. Infrared reflection spectrum of an SOG layer baked at 150°C.

The residual organic radicals remained from the solvent in the SOG layer can easily hydrolyze with the water molecules adsorbed at the Si surface and create strong covalent bonds. Even after a baking procedure at 150°C for 5 minutes the SOG surface contains also silanol groups, which may generate adhesion at RT due to their ability to form bonds with molecules from the other surface directly or *via* water molecules.

After RT bonding the samples were annealed at temperatures in the range 200°C –300°C for different tims. IR transmission images and acoustic micrographs (figure 2) showed good quality interfaces after the thermal annealing.

Figure 2. Acoustic micrograph of a bonded Si/GaAs wafers pair.
Grey contrast shows the bonded interface, white spots represent unbonded areas.

The surface energy increased up to 2 J/m^2 after an annealing at 200°C for 10 hours. Further increasing of the annealing temperature does not produce a significant increase of the surface energy. Some of the Si/GaAs bonded wafer pairs were heated up to higher temperatures. At about 280°C the wafers were debonding and shattered.

Samples of 6 x 6 mm^2 were cut from the bonded pairs annealed at 200°C and submitted to tensile stress testing. The samples cracked at values of about 22 MPa. AFM (figure 3) and SEM (not presented here) investigation of the resulted broken surfaces revealed that the SOG film brakes during the tensile test. This demonstrates the suitability of the SOG layer as "adhesion layer".

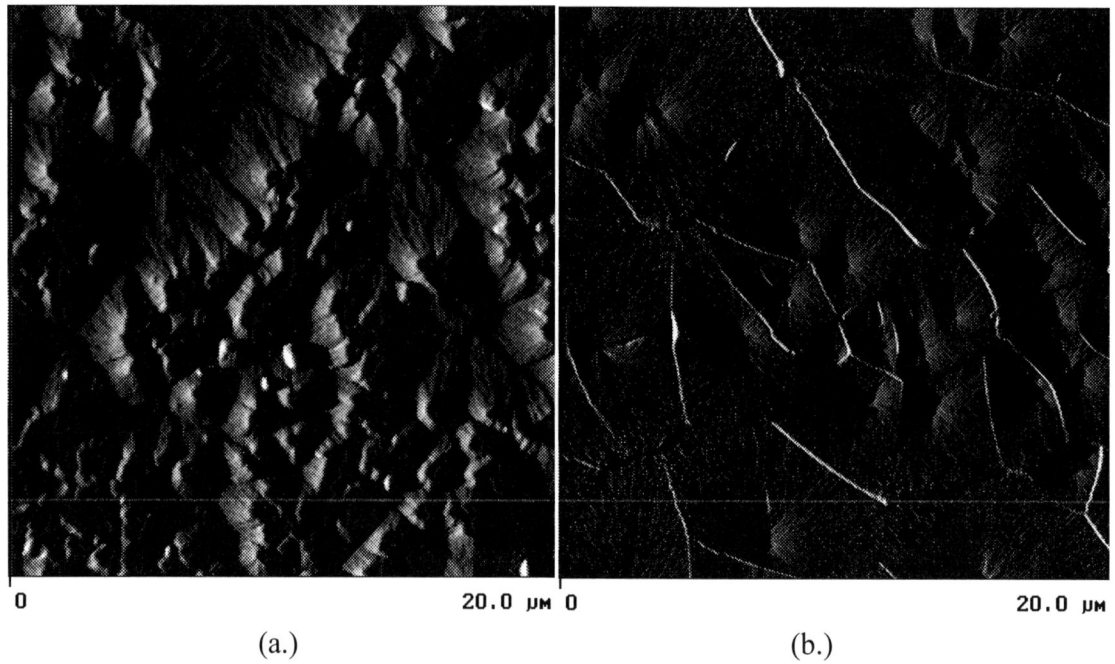

(a.)   (b.)

Figure 3. AFM images of (a.)- Si wafer, and (b.)-GaAs wafer
surfaces resulted after tensile esting.

The bow of the bonded wafer pairs was measured *in situ* during annealing. At 200°C a high bow of about 500 μm was measured. The bow is decreasing to zero during cooling of the bonded pair down to RT. For repeated heating-cooling cycles with the same bonded pair it was observed that the bow follows the same path at heating as well as cooling, showing no hysteresis. SAM investigations revealed that the bonded interface remained intact during this test, no voids being generated. This elastic behavior of the SOG layer compensates the high stress developed at the interface due to the thermal mismatch.

In order to prove that the surface energy achieved after annealing at 200°C is high enough to allow the subsequent processing of the as obtained Si/GaAs heterostructures, one of the bonded wafers was thinned by grinding followed by chemical mechanical polishing (CMP). The thinning procedure was applied to GaAs (thinned down to about 10 μm) as well as to Si (thinned down to about 5 μm). After thinning, the GaAs-on-Si and Si-on-GaAs resulting wafers were heated at 450°C. Further investigations revealed that the bonded interface remained unchanged, no voids being generated. The thinning of one of the bonding partners diminish the thermally induced stress which causes the debonding of the thick wafers.

APPLICATIONS OF BONDING WITH SOG INTERMEDIATE LAYERS

The SOG bonding procedure described above was developed as an industrial technology for producing universal GaAs-on-Si substrates. By this method, substrates having different insulator or GaAs layer thickness can be easily produced for wafer diameter up to 150 mm.

The same bonding approach was applied for bonding GaAs epitaxial wafers with CMOS Si wafers. In this case, the main problem encountered is the planarization of the structured CMOS wafer. SOG is widely used for planarization in microelectronics industry but is not efficient when large area structures are present on the surface (in this case, about 5 x 5 mm^2 with 1.5 μm depth). Figure 4 shows the profile of a structure measured with a profiler.

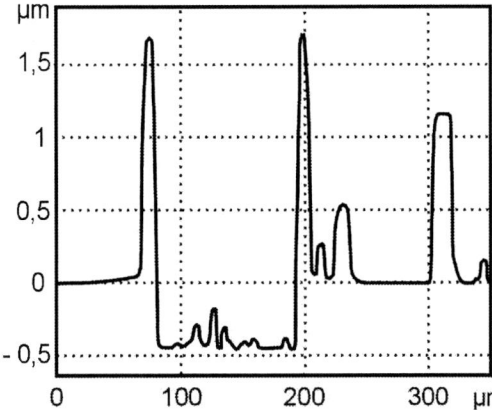

Figure 4. Profile of a structure on the CMOS Si wafer surface.

An efficient planarization method was the deposition of a 1 μm thick Si_3N_4 layer on top of the structured wafer, followed by the deposition of a 2 μm thick SiO_2 layer. The oxide was then planarized using CMP. The nitride layer acts as a polish stop layer. After the CMP planarization an SOG layer was deposited for a final planarization and as a bonding intermediate layer. The planarized wafers were bonded with GaAs wafers as described. The low processing temperature prevents both damage of the structures and decomposition of the compound semiconductor. Using the above described process, GaAs multiple quantum wells (MQW) were transferred to a standard CMOS Si wafer. MQW multilayer structures were grown on GaAs substrate and then bonded to a planarized CMOS Si wafer. Finally the GaAs substrate was back etched and the multilayer structures were patterned.

CONCLUSIONS

A new method was successfully applied for Si/GaAs heterostructures fabrication: bonding through an intermediate spin-on glass layer. The main advantages of SOG use are: i) SOG is an electronically clean material, which does not produce contamination of Si or GaAS and ii) SOG can be easily deposited at RT using an usual spinner.

The surface energy for Si/GaAs pairs bonded with this procedure is about 0.4 J/m^2 for RT bonded wafers and increases to about 2 J/m^2 for annealing at 200°C for 10 hours.

The thermally induced stress develops a high bow at 200°C (about 500 µm) and leads to debonding at temperatures close to 300°C.

Si-on-GaAs (6 µm Si top layer thickness) and GaAs-on-Si (10 µm GaAs top layer thickness) wafers were fabricated by thinning one of the bonded wafers by grinding and CMP. The as obtained wafers sustain temperatures of 450°C without damaging of the bonded interface.

This bonding process was applied for bonding GaAs epitaxial wafers to CMOS Si wafers. The CMOS Si wafer was planarized by deposition of a polish stop Si_3N_4 layer and a SiO_2 layer followed by CMP. The electrical interconnections between the CMOS structures and the top GaAs layer can be realized by etching vias into the insulator layer.

ACKNOWLEDGEMENTS

The authors would like to thank to S. Hopfe, M. Wiegand and K. P. Meyer for their valuable help. This research was supported by ESPRIT-BONTEC Project under contract no. 28998 and by the Thüringen Ministry for Science, Research and Culture under contract no. B609-97036.

REFERENCES

1. P. J. Taylor, W. A. Jesser, J. D. Benson, M. Martinka, J. H. Dinan, J. Bradshaw, M. Lara-Taysing, R. P. Leavitt, G. Simonis, W. Chang, W. W. Clarck III and K. A. Bertness, *J. of Appl. Phys.* **89**, 4365 (2001).
2. Q. Y. Tong and U. Gösele, "Semiconductor Wafer Bonding – Science and Technology", John Wiley & Sons, New York, 1999.
3. N. Chand in "Properties of Gallium Arsenide" (2nd ed.), EMIS Datareviews Series no. 2, 1990, pp. 459.
4. Q.-Y. Tong, W. J. Kim, T.-H. Lee and U. Gösele, *Electrochem. Solid-State Lett.* **1**, 52 (1998).
5. S. N. Farrens, J. R. Dekker, J. K. Smith and B. E. Roberds, *J. of the Electrochem. Soc.* **142**, 3950 (1995)
6. M. Wiegand, M. Reiche and U. Gösele, *J. of the Electrochem. Soc.* **147**, 2734 (2000)
7. G. Kräuter, A. Schumacher U. Gösele, T. Jaworek and G. Wegner, *Adv. Mater.* **9**, 417 (1997).
8. D. M. Hansen, P. D. Moran, K. A. Dunn, S. E. Babcock, R. J. Matiy and T. F. Kuech, *J. of Chryst. Growth* **195**, 144 (1998).
9. M. Alexe, V. Dragoi, M. Reiche and U. Gösele, *Electronic Lett.* **36**, 677 (2000).

Mat. Res. Soc. Symp. Proc. Vol. 681E © 2001 Materials Research Society

SiC-Si Grooved Surface Bonding

Tatiana S. Agrunova, Igor V. Grekhov, Lioudmila S. Kostina, Alexander G. Tur'yanskii[1], Igor V. Pirshin[1], Ilya R. Prudnikov[2] and Konstantin B. Kostin[3]

Solid State Electronics Division, Ioffe Physico-Technical Institute RAS,
Polytekhnicheskaya ul. 26, 194021, St. Petersburg, RUSSIA
[1]Lebedev Physical Institute RAS, Leninskii pr. 53, 117924, Moscow, RUSSIA
[2]Physics Department, Moscow State University, Vorob'evy Gory, 119899, Moscow, RUSSIA
[3]Materials Science Department, Christian-Albrechts-University of Kiel,
Kaiserstr. 2, Kiel 24143, GERMANY

ABSTRACT

SiC Lely platelets and SiC epilayers on large area SiC substrates were directly bonded to non-oxidized grooved surface silicon wafers in order to obtain structures prospective for the design of power bipolar devices with wide band-gap emitter junctions. The reported capabilities of grooved interfaces to reduce elastic strain and intrinsic sources of an interfacial potential barrier were utilized. The influence of surface morphology on the structural perfection of the bonded samples was studied in detail by X-ray and AFM techniques. As compared to traditional bonding technology, experimental data showed an easier smooth-to-grooved surface bonding accomplished in the formation of the boundary with better continuity and strength. For 2 in. diameter SiC wafers with root mean square height of roughness $\sigma=16$ Å and lateral coherence length $L=1.8$ μm bonding continuity not smaller than 90% was reached, while the crystals with $\sigma=30$ Å, $L=5$ μm failed to bond to Si even under an external force.

INTRODUCTION

Silicon carbide and compositions on its base due to their unique properties are being used for the design of high temperature circuits, power high voltage bipolar and field effect devices *etc*. A wafer bonding approach to fabricate SiC layers on large area insulating substrates was advanced in [1]. Si-SiO_2- SiC compositions were made in such a way that β-SiC layers epitaxially grown on Si substrates were transferred onto oxidized Si wafers by bonding and etchback. Though successful as whole, the technology appeared to have shortcomings caused in particular by roughness of SiC surface. Bonding strengthening via multistep annealing between etches of Si substrate was proposed. An increasing flexibility of thinned SiC-Si portion of Si-SiO_2-(SiC-Si) bonded structure was thus exploited to enhance the bonding energy.

In this paper we demonstrate the other way by which the influence of the surface roughness can be appreciably reduced. The communication deals with structures prospective for the design of power bipolar devices with wide band-gap emitter junctions. 6H SiC wafers 0.3÷0.4 mm thick made of Lely grown crystals or composed by CVD layers on SiC substrates were directly bonded to silicon in order to provide oxide-free SiC-Si bonded compositions. It was earlier shown that, in the course of Si-Si bonding, an interfacial grooved relief eliminated elastic strain caused by surface morphology [2]. Besides, during annealing, the grooves acted as sinks for dislocations reducing intrinsic sources of an interfacial potential barrier [3]. In the studied structures the grooved relief was made on Si wafer. Quality of smooth and grooved SiC-Si bonded interfaces was compared. The influence of surface morphology on the structural perfection of the bonded samples was investigated in detail.

EXPERIMENTAL DETAILS

Lely platelets were mirror-like mechanically polished on rigid pads and atmospheric pressure hydrogen etched at 1550°C. CVD layers of thickness 5 μm on 0.3 mm thick 6H SiC substrates were produced by Cree Inc. SiC impurity concentration was ~5·10[18] cm[-3].

Roughness characteristics of SiC surface, root mean square (RMS) height σ and lateral coherence length L, were determined from X-ray specular spectra and diffuse scattering under specular conditions. The device was a two-wave reflectometer that provided simultaneous measurements of scattering diagrams and angle dependencies of reflectivity in two spectral regions (here CuK_α and K_β) [4]. The basic advantage of this design was the possibility of performing relative measurements. This allowed one to eliminate or significantly reduce a number of principal errors of conventional one-wave reflectometry: (i) the dynamic measurement range could be significantly extended; (ii) the operating range of scanning angles was also extended; (iii) limitations for the linear dimensions of the measured sample were cancelled. For the simulation of the profiles different models were utilized. The extraction of the contributions caused by curvature and surface roughness gave an opportunity to compare the calculated σ and L values for two wavelengths. AFM technique was the other tool used for the evaluation of SiC surface quality.

Orthogonal cross-hatched grooved relief was made on the surface of Si wafers by a photolithography technique. The groove width, depth and spacing were 50, 0.3 and 200 μm respectively. The wafers to be bonded were subjected to RCA cleaning. (0001) on-axis Si-face SiC wafers were bonded to (111) Si ones. The bonding was realized in the air at temperature elevated from room to 1150°C. The continuity of the interface was studied by X-ray diffraction topography in Laue geometry. Fracture strength was assessed by cutting the samples.

DISCUSSION

X-ray topography data showed an easier smooth-to-grooved surface bonding accomplished in the formation of the boundary with better continuity. The result was attributed to an appreciable roughness of SiC surface, which was evaluated as follows.

The reflectivity angular dependence $R(\theta)$ measured in the $\theta-2\theta$ scanning mode is sensitive to RMS height of roughness σ and the depth distribution of electron density ρ averaged over sample surface. These structural parameters were determined by the simulation of specular spectra. In this work numerical simulation was done by using Fresnel reflection coefficient multiplied by statistical Debye-Waller factor [5]. Figure 1 demonstrates the experimental reflectogram and the result of its theoretical treatment for the surface of SiC CVD layer on SiC substrate. RMS heights of roughness were found to be equal: σ_1=(16±2) Å, σ_2=(3.2±0.3) Å, and the density ρ=2.8 g/cm[3].

The density of oxide layer on SiC surface as well as the SiC density was measured by X-ray refraction technique performed on the basis of the experimental set up used in the present work. The obtained values were as follows: ρ_{oxide}=2.1 g/cm[3], ρ_{SiC}=3.0 g/cm[3]. Thus a reduced density value ρ=2.8 g/cm[3] derived from fitting of $\theta-2\theta$ reflectivity curve (Figure 1) could be interpreted due to the presence of silicon oxide layer on SiC surface.

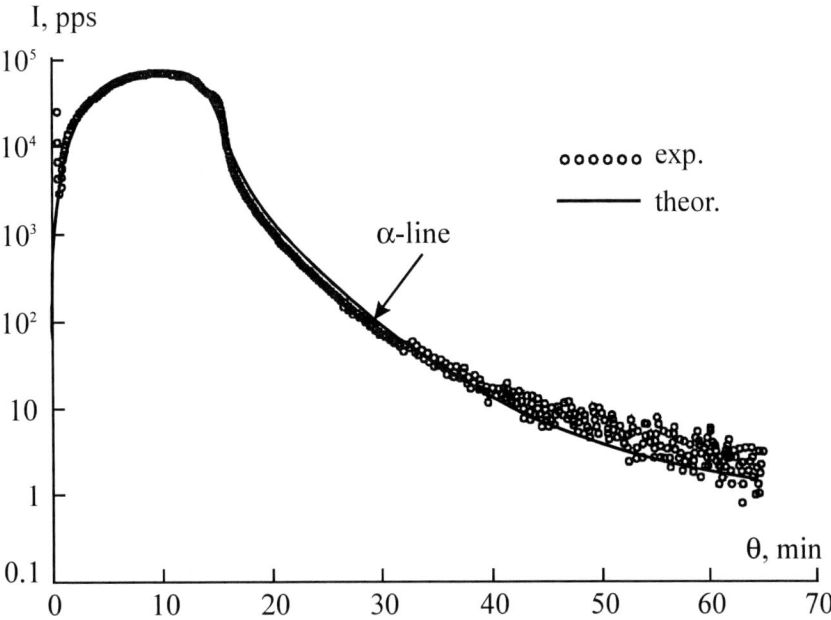

Figure 1. Angular dependence of the intensity reflected from SiC surface in θ-2θ scanning mode and the result of its simulation. The sample is 6H CVD epilayer.

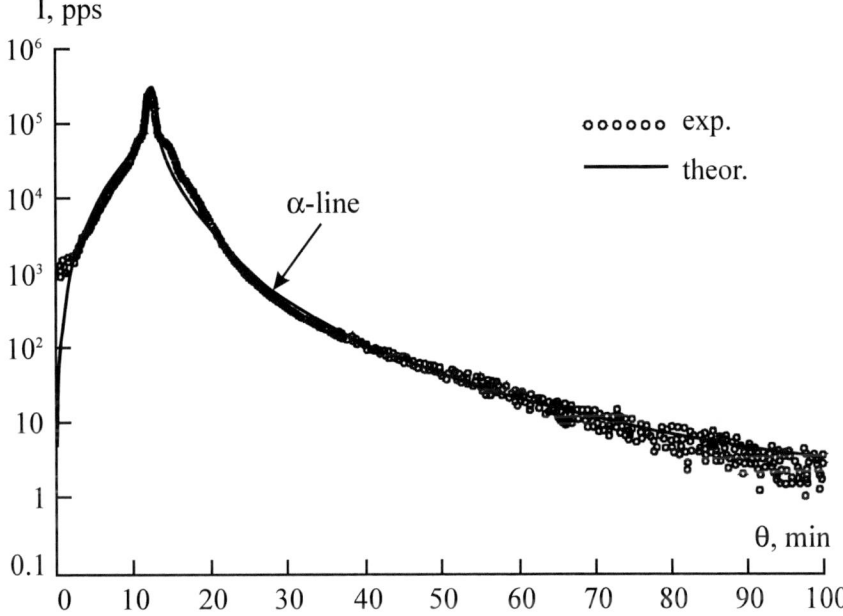

Figure 2. Angular distribution of scattered intensity from SiC surface (CVD epilayer) measured by 2θ-scan and theoretical curve. Incidence angle of X-ray beam $\theta_0 = 11'$.

The simulation of X-ray diffuse scattering measured under specular conditions allows establishing the RMS height of roughness σ as well as lateral coherence length L. The diffuse scattering intensity was calculated by using the method of Born approximation of distorted waves [6, 7]. The exponential correlation function was utilized [7]. The predetermined σ and ρ values were used for fitting. The results of simulation of 2θ-scan scattering diagrams and θ-scan rocking curves are shown in Figures 2, 3. The lateral coherence lengths were determined

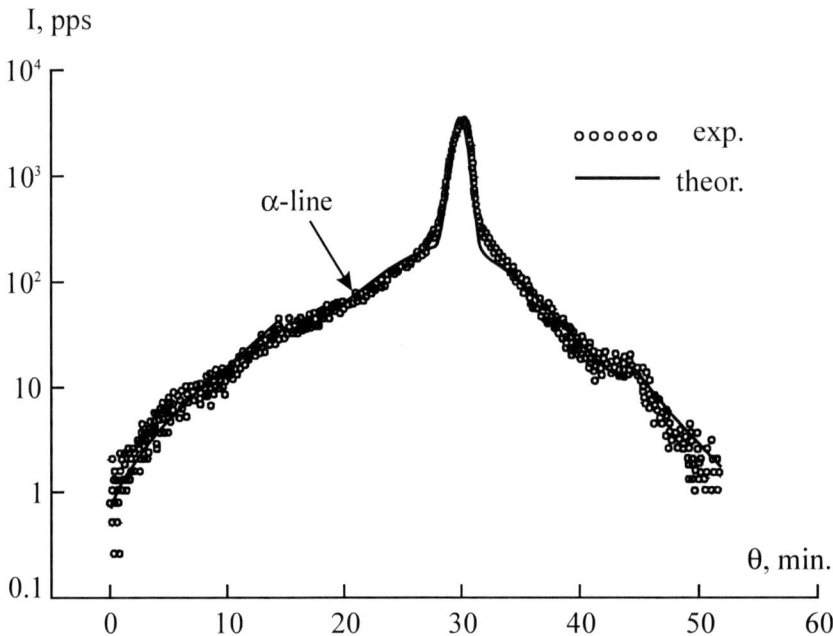

Figure 3. Intensity distribution in θ-scan from the surface of SiC CVD layer and the result of the simulation of the rocking curve.

to be: L_1=1.8 μm (σ_1=16 Å) and L_2=0.6 μm (σ_2=3.2 Å). Table 1 contains the X-ray data representative for a series of measurements. The results are displayed in comparison with AFM measurements.

Figure 4 demonstrates AFM pictures of the surface of the CVD layer described above. Processing of scans allowed to conclude that white spots (Figure 4*a*) were characterized by RMS height of roughness σ_1=15 Å and L_1=5 μm. Circular-like relief (Figure 4*b*) gave: σ_2=3.1 Å, L_2=0.7 μm. The third roughness component, not detected by X-ray reflectometry technique, was characterized by σ_3=1.8 Å and L_3=200 Å.

Table 1. Surface roughness of SiC samples

Sample	Technique	Roughness characteristics: RMS height of roughness, σ; lateral coherence length, L		
		σ_1, Å; L_1, μm	σ_2, Å; L_2, μm	σ_3, Å; L_3, Å
6H CVD epilayer	X-ray reflectometry	(16±2); 1.8	(3.2±0.3); 0.6	
	AFM	15; 5	3.1; 0.7	1.8; 200
6H Lely	X-ray reflectometry	(30±5); 5		

The comparison of the results shows a complementary character of AFM and X-ray data. From the Table 1 it is seen that the discrepancy between the coherence lengths determined by two techniques is appreciable for bigger L values, but it remarkably diminishes with L decreasing. Since the AFM view field does not exceed 10×10 μm^2, only few "peaks" 15 Å in height (σ_1) can be detected. The parameters obtained from the simulation of X-ray spectra result

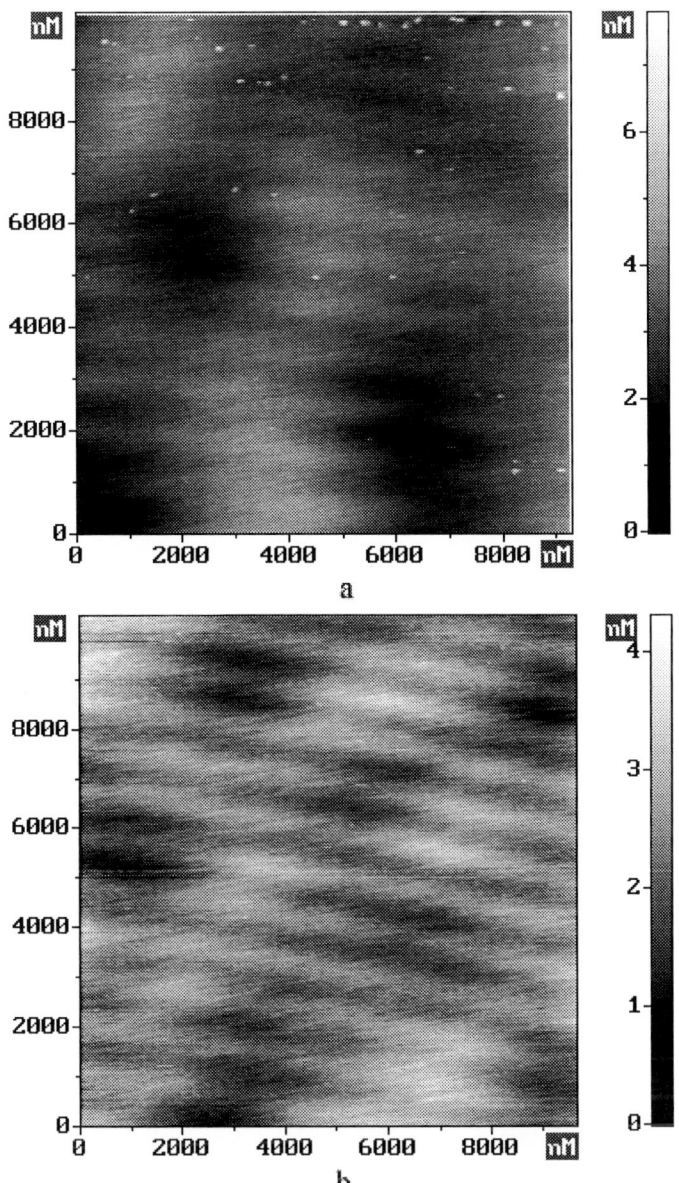

Figure 4. AFM pictures of SiC surface (CVD layer). (a) and (b) images show the 1st and the 2nd order roughness components correspondingly: a-σ_1=15 Å; L_1=5 µm; b-σ_2=3.1 Å; L_2=0.7 µm.

from averaging over at least few centimeters of sample area and, for the roughness with long coherence length values, they are more reliable. A good agreement is observed for "medium" σ_2, L_2 roughness components. For the detection of a small-height roughness AFM method is certainly advantageous because of low X-ray intensities reflected at bigger grazing angles.

For the surface of Lely platelets the simulation of X-ray reflectivity and diffuse scattering spectra gave: σ=(30±5) Å, L=5 µm. The obtained values did not disagree with AFM data. Though the height of roughness for this sample appeared to be bigger than before, its surface was

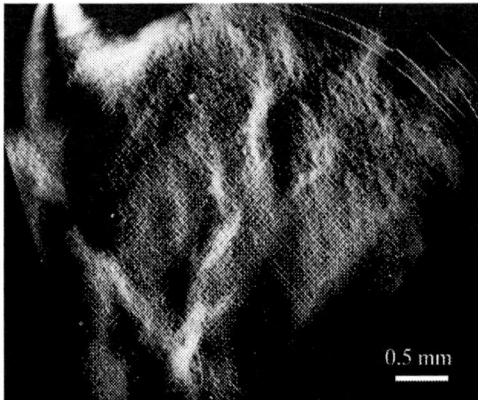

0.5 mm

Figure 5. X-ray topograph of SiC(subst.)/SiC(layer)-Si composition with SiC(layer)-Si bonding interface. Lang technique. 022, MoK_α.

still flat since σ/L ratio was far less than unity: $\sigma/L \sim 10^{-3} \ll 1$.

By grooved surface bonding technology SiC CVD layers could be bonded to silicon with continuity not smaller than 90%. Figure 5 shows a typical X-ray topograph of such a structure. Grooved interface and SiC-Si interfacial dislocations of high density are seen. The attempts to bond SiC bulk crystals (Lely platelets) were not successful. However, we believe that for the bulk crystals better surface smoothness is reachable and their bonding to non-oxidized Si wafers is principally possible.

CONCLUSIONS

Large area SiC substrates covered by SiC epilayers with RMS height of roughness ~16 Å and lateral coherence length 1.8 μm can be directly bonded to silicon via grooved surface bonding technology to provide a wide band-gap emitter junction. For 2 in. diameter SiC wafers interface continuity not smaller than 90% can be reached. Due to a complementary character of X-ray reflectometry and AFM techniques a comprehensive characterization of surfaces to be bonded can be obtained. By choosing proper parameters of silicon surface grooved relief strain coming from a non-flat SiC surface can be reduced.

REFERENCES

1. Q-Y. Tong, U. Gösele, C. Yuan, A. Steckl, and M. Reiche, *J. Electroc. Soc.* **142**, 232 (1995).
2. T. Argunova, R. Vitman, I. Grekhov, L. Kostina, T. Kudryavtseva, M. Gutkin, A. Shturbin, J. Härtwig, M. Ohler, E. Kim, and S. Kim, *Phys. Solid State* **41**, 1790 (1999).
3. T. Argunova, I. Grekhov, M. Gutkin, L. Kostina, E. Belyakova, T. Kudryavtseva, E. Kim, and D. Park, *Phys. Solid State* **38**, 1832 (1996).
4. A. Tur'yanski, A. Vinogradov, and I. Pirshin. *Instrum.&Exper. Tech.*, **42**, 94 (1999).
5. L. Nevot, P. Croce. *Rev. Phys. Appl.* **15**, 761 (1980).
6. A. Andreev, A. Michette and A. Renwick *J. Mod. Opt.* **35** (1988) 1667
7. S. Sinha, E. Sirota, S. Garoff, H. Stanley *Phys. Rev. B.* **38**, 2297 (1988).

Mat. Res. Soc. Symp. Proc. Vol. 681E © 2001 Materials Research Society

Direct Bonding of Silicon Wafers with Simultaneous Dopant Diffusion

Igor V. Grekhov, Tatiana S. Agrunova, Lioudmila S. Kostina, Natalia M. Shmidt, Helmut Föll[1], and Konstantin B. Kostin[1]

Solid State Electronics Division, Ioffe Physico-Technical Institute RAS, Polytekhnicheskaya ul. 26, 194021, St. Petersburg, RUSSIA

[1]Materials Science Department, Christian-Albrechts-University of Kiel, Kaiserstr. 2, Kiel 24143, GERMANY

ABSTRACT

Bonding of silicon surfaces in aqueous solution of compounds containing III and IV impurities was performed for the first time. It was observed that the presence of aluminum at the bonding interface improved structural quality of the interface. This phenomenon is explained by the increase of the contact area due to Al-OH group sandwiched between the water molecules adsorbed at hydrophilic wafer surfaces at the first bonding stage. The incorporation of Al produces a p-type layer and the I/V characteristics of the resultant np^+n diodes is shown not to be influenced by the presence of the bonding interface. The technique developed could be advantageous for the design of multi-layer large area semiconductor devices.

INTRODUCTION

Direct bonding technology can be used to bond partially or fully processed wafers to desired substrates [1,2]. However, doping selected areas or heavy doping of the whole wafer as required for device manufacturing before bonding often results in a deterioration of the wafer surface quality as well as in elastic strain. Bonding than may be difficult and imperfect or impossible. Large area bonding with an oxide free interface under those conditions remains a challenge up to now. In a monograph [1] and two reviews [2,3] different aspects of this problem– bonding conditions, reasons of bubble appearance, real surface structure influence–are discussed in detail. To decrease the detrimental influence of elastic strain and an imperfect surface morphology on bonding and the structural quality of the bonding interface, a grooved-surface direct bonding technology was suggested [4,5].

In this paper, the formation of p- or n-type layers in silicon wafers is proposed to be performed directly in the course of the direct bonding procedure by supplying a diffusion source placed between the wafers to be bonded. The structural and electronic quality of the resulting structural and electronic junctions was investigated and found to be very good for Al-based diffusion source. A model is suggested to explain the high interface continuity in the structures with interfacial aluminum.

EXPERIMENTAL DETAILS

Commercially available mirror-polished float-zone (FZ) n- and p- type silicon wafers of ~20 Ω·cm as well as p-type Czochralski (CZ) wafers of 0.005 Ω·cm, both 40-60 mm in diameter and (111) orientation, were used. An orthogonal net of grooves with a depth 0.2-0.3 μm, a width 50 μm and spaced at 200 μm, was made by photolithography on the surface of one wafer from each pair before bonding. The bonding procedure included the following steps: 1) the wafer

RCA cleaning; 2) the wafer attachment either in DI water of 18 MΩ·cm or in 0.1-0.5wt% aqueous solutions of $Al(NO_3)_3$, $Ga(NO_3)_3$, H_3PO_4 or HBO_3 used as diffusion sources; 3) spin dry of the attached wafers; 4) annealing in air successively at 95^0C for 4 hours followed by 1250^0C for 4 hours. No efforts were made for aligning the wafer crystal lattices, so a random grain boundary must be expected.

The structural quality of the bonded wafers was investigated by X-ray topography (XRT) and SEM techniques. The impurity diffusion profile and diode I/V characteristic were measured using spreading resistance and a pulsed high power I/V measurement system, respectively.

DISCUSSION

Careful XRT and SEM examination showed that the initial wafer mating in the aqueous solutions containing such impurities as gallium, boron and phosphorous did neither result in deterioration nor in a significant improvement of the bonding interface quality in comparison to clean DI water bonding. The net of grooves, as well as occasional bubbles could be observed at the boundary after the bonding procedure was completed.

Bonded structures made via wafer attachment in aluminum nitrate solution, however, showed a substantially higher structural quality of the interface. Cross-sectional SEM images (Figure 1) do not reveal the interface any more which indicates that the grooves are filled.

Figures 2*a,b* present typical X-ray topographs of "Al bonded" compositions. Over a whole wafer area no bubbles are seen. The images still show a periodic contrast modulation preserving the geometry of the grooved lattice together with a waviness-like contrast arising from non-flatness of the wafer surfaces. These contrasts result from elastic strain necessary to make the grooved and non-flat surfaces adhere strongly [6]. The absence of such contrasts indicates absence of bonding (Figures 2*c*), and the extent of contrast free regions may be used as a measure of bonding quality. A quantitative evaluation relied on computer processing (via "sharpen edge" filters and conversion of the grayscale image to a bitmap mode) followed by the calculation of the disbonded area.

The results for 56 "DI water bonded" and for 37 "Al bonded" samples are presented in Figure 3 as histograms. While the bonded interface fraction for "Al bonded" compositions is close to 100% the "DI water bonded" samples peak at ~90%.

For the explanation of the results obtained, a model of room-temperature bonding of hydrophilic surfaces as presented in Figure 4 was adopted [7]. With a linkage of adsorbed water molecules of ~7 Å, bonding occurs over wafer surface areas with microroughness smaller than

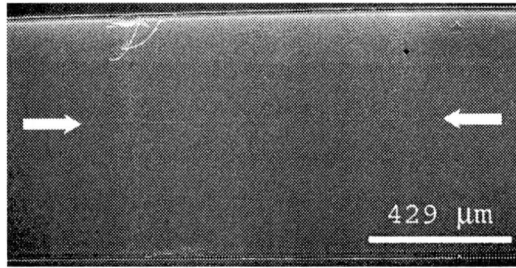

Figure 1. SEM cross-section view of Si samples bonded using aqueous aluminum nitrate solution

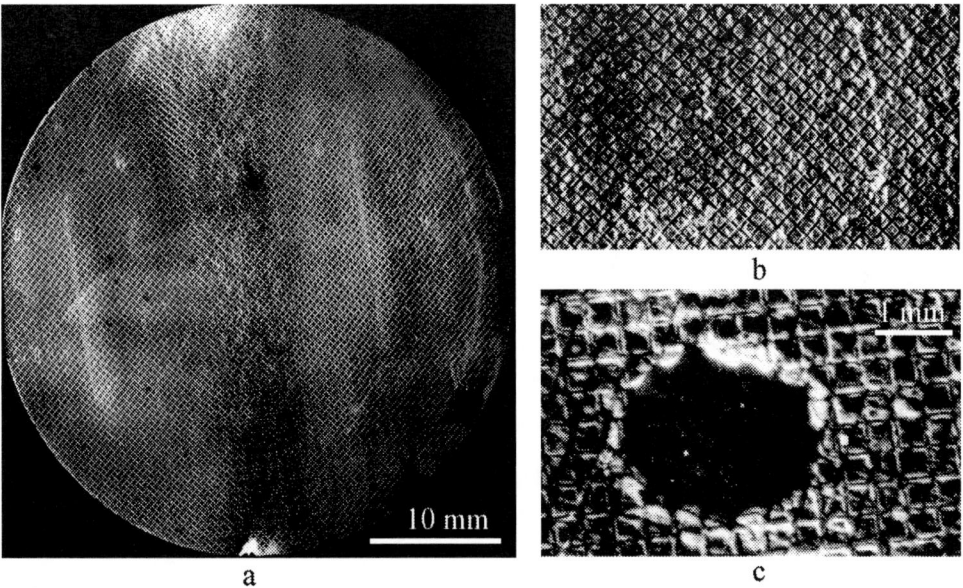

Figure 2. *a,b*-typical X-ray topographs of "Al-bonded" compositions; *c*-XRT image of the disbonded area. Lang technique, 220 MoK_α.

3.5 Å. At ~150°C, almost all silanol Si-OH groups in the contacted area are converted to strong siloxane Si-O-Si bonds [1], the action radius of which, however, is considerably shorter (1.6 Å [7]). This means that interface continuity could be achieved if the contacted areas conform in their roughness to the 7 Å "window" provided by hydrogen bonds.

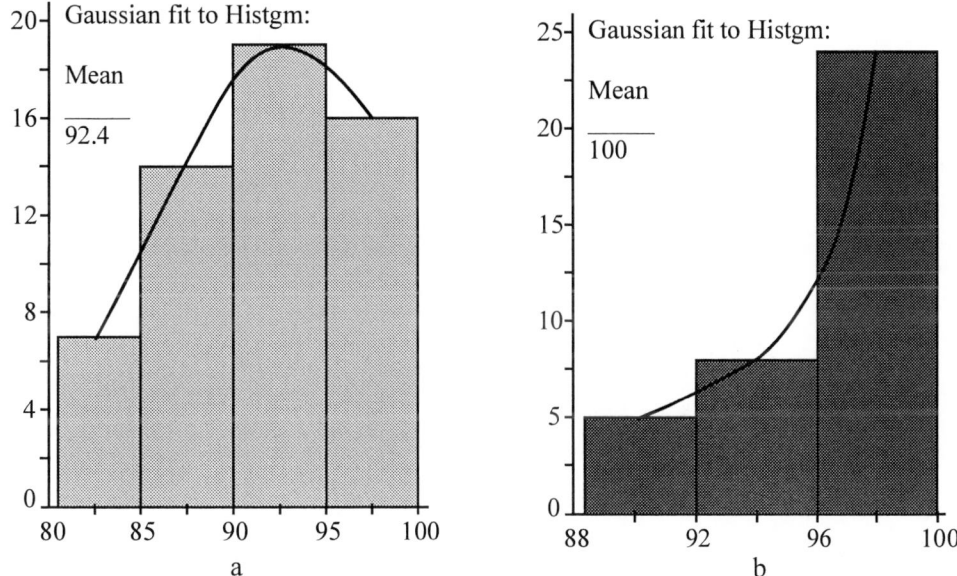

Figure 3. Histograms of the fraction of bonded areas: a) "DI water bonded" and b) "Al-bonded"

In the case of "Al-bonded" samples, the continuous interface bonding may have a larger roughness window in the following way. A dissociation of aluminum nitrate molecules in the aqueous solution

$$Al(NO_3)_3 => Al^{3+} + 3NO_3 \qquad (1)$$

results in the interaction of Al^{3+} ions with water molecules, and in the creation of aluminum hydroxide $Al(OH)_3$. Due to Al-OH group built-in into OH-OH "bridge" between the wafers, its length can be increased, as it is shown in Figure 4. As a consequence, surface areas with a microroughness larger than 3.5 Å could be incorporated into the hydrogen bonding and a more extended contact area results. With temperature increase, hydrogen bonds are replaced with stronger siloxane bonds:

$$Si\text{-}OH\text{-}2H_2O\text{-}2Al(OH)_3\text{-}2H_2O\text{-}OH\text{-}Si => Si\text{-}O\text{-}Al\text{-}Al\text{-}Si + 8H_2O \qquad (2)$$
$$| \quad |$$
$$O \quad O$$

At the elevated temperatures the aluminum atoms could react in two ways. On the one hand, they could reduce the native oxide via [8]

$$(4/3)Al + SiO_2 = (2/3)Al_2O_3 + Si \qquad (3)$$

which results in the release of free Si atoms that are capable to participate in the final stage of the bonding process, especially by filling possible microgaps via surface diffusion mechanism. On the other hand, because the Al-Si bond energy is higher than Al-Al one, introduction of

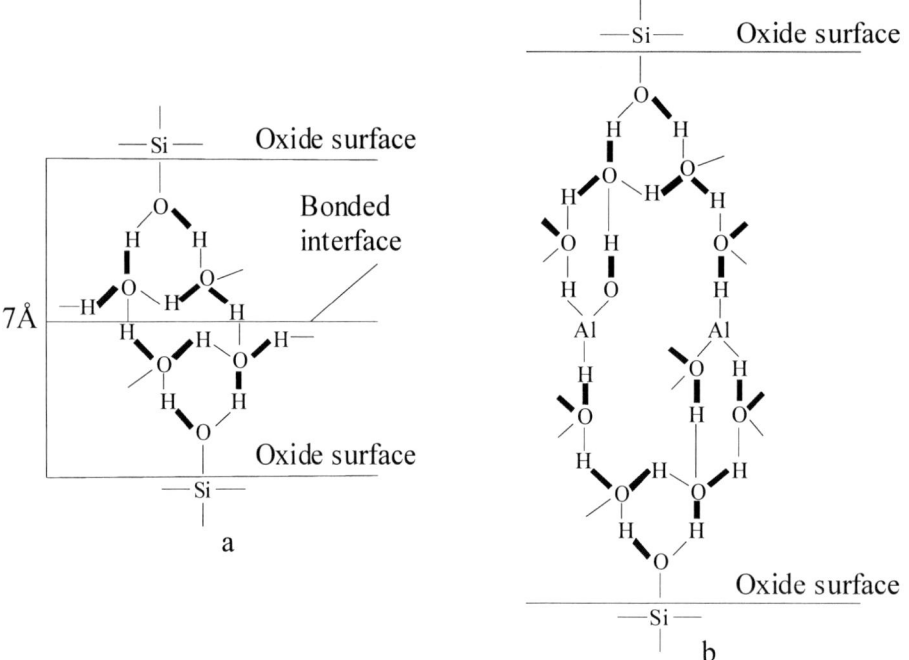

Figure 4. Model of hydrophilic bonding: a) "DI water-bonded" [7]; b) "Al-bonded" samples

aluminum atoms into the silicon lattice produces Si interstitials and stimulates the self-diffusion of Si atoms, again promoting microgap closing.

In any case, some Al will diffuse into the bonded wafers and produce a p-doped layer. An investigation of the aluminum diffusion profile in n/n-bonded structures is shown in Figure 5. It is clear that an extended p-type diffusion layer can be formed by aluminum diffusion into the polished silicon surface from a chemically deposited source at the bonding interface in the course of the direct bonding procedure during the high temperature annealing in air. Under the chosen annealing conditions, the total diffusion layer thickness is about 30 μm with an Al concentration at the bonding interface of ~5·10^{16} cm^{-3}.

In order to investigate the electrical properties of the pn-junctions formed by bonding and Al diffusion, p$^+$n diodes were made by direct bonding of a p$^+$ type 0.005 Ω·cm silicon wafer to a n-type 20 Ω·cm wafer via their mating in 0.5wt% Al(NO$_3$)$_3$ aqueous solution, with the Al diffusion providing a p layer extending somewhat into the n-type Si. Figure 6 presents the I(V) characteristics of the junction. The forward I(V) curve was measured in a pulse mode with a pulse duration of 30 μs. Owing to the deep aluminum into the bulk n-type wafer, the bonding interface in the p-doped part of the wafers (necessarily containing a grain boundary) should not influence the diode characteristics, which is confirmed by the experiment. The I(V) curve is typical for a diode made by conventional diffusion with a forward voltage drop about 2.0 V at a current density ~200 A/cm^2. The diode breakdown voltage is about 1200 V, which corresponds to the chosen n-type silicon resistivity.

CONCLUSIONS

This study shows for the first time that high quality multi-layer semiconductor structures with p-n junctions could be made via direct bonding and the incorporation of a Al diffusion source. A substantial improvement of the bonding interface quality in sandwiches with

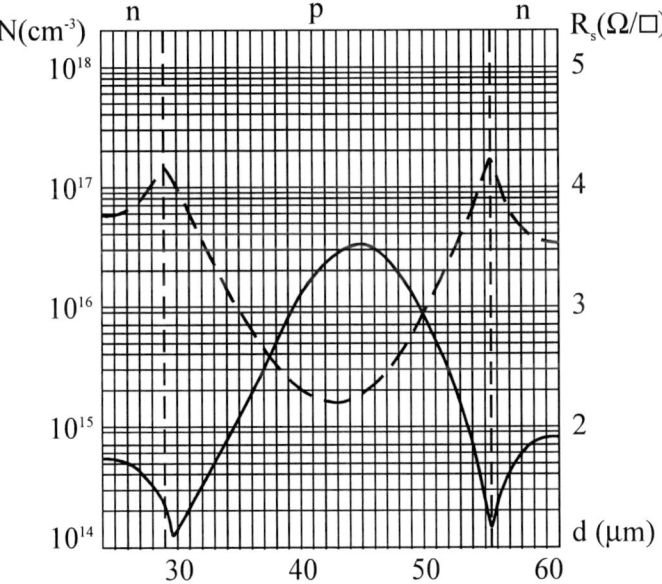

Figure 5. Spreading resistance profile (dashed) and the corresponding Al diffusion profile (solid).

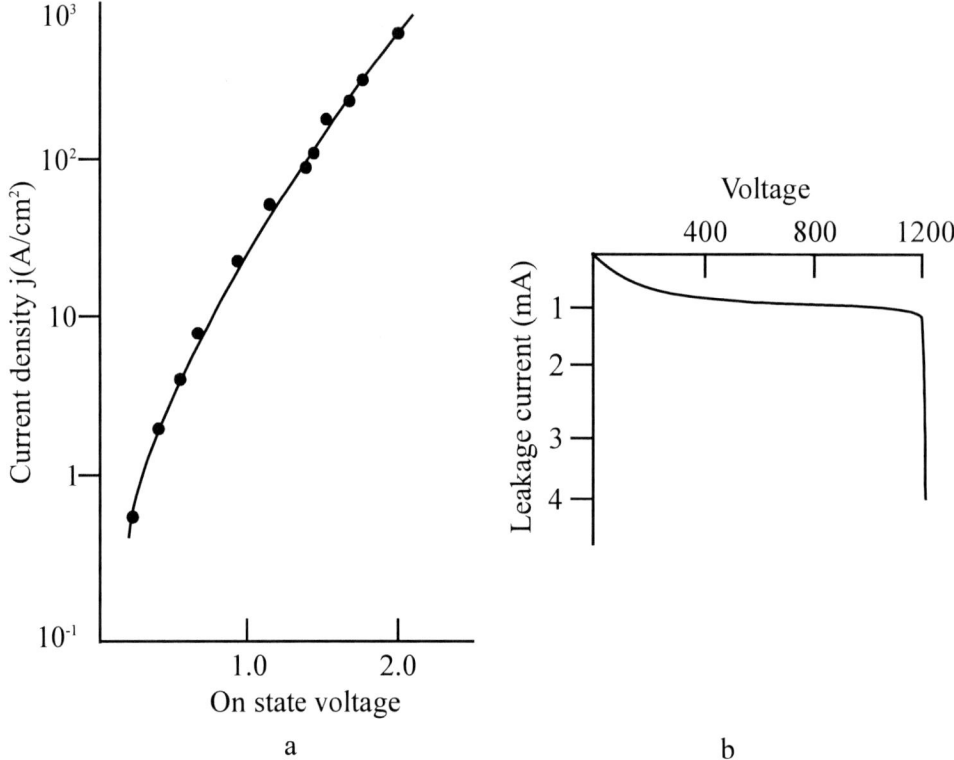

Figure 6. Current-voltage characteristic of p^+n-"Al-bonded"diode with an operational area S~4 cm^2.

interfacial aluminum was attributed to the formation of Al-OH "bridging" between the water molecules adsorbed on the contacted wafers. The I/V characteristics of diodes formed in this way confirm the potential of the proposed technique for the production of large area semiconductor devices with pn-junctions.

REFERENCES

1. Q.-Y. Tong, U. Gösele , *Semiconductor Wafer Bonding: Science and Technology,*
(J. Wiley&Sons, Inc., 1999) pp.49-123.
2. S. Bengtsson, *J. of Electronic Mater.* **21**, 841 (1992).
3. A. Plößl and G. Kräuter, Mater. Sci.&Eng. **R25**, 1 (1999).
4. E.D. Kim, N.K. Kim. S.C. Kim, I.V. Grekhov, T.S. Argunova, L.S. Kostina, and
T.V. Kudriavtseva, *Electronic Letters* **31**, 2047 (1995).
5. T.S. Argunova, R.F. Vitman, I.V. Grekhov, M.Yu Gutkin, L.S. Kostina, A.V. Shturbin,
J. Häertwig, M. Ohler, E.D. Kim, S.C. Kim, *Solid State Physics* **41**, 1790 (1999).
6. W.P. Maszara, B.-L. Jiang, A. Yamada, G.A. Rosgonyi, H. Baumgart, and A.J.R. de Kock,
J. Appl. Phys. **69**, 257 (1991).
7. R. Stengl, T. Tan, and U. Gösele, , *Jpn. J. Appl. Phys.* **28**, 1735 (1989).
8. M. Ohring, *The Materials Science and Thin Films,* (Acad. Press Limited, 1991) p.493.

Mat. Res. Soc. Symp. Proc. Vol. 681E © 2001 Materials Research Society

Thermomechanical stress in silicon on quartz wafer bonding and Smart Cut® process

Yu-Lin Chao,[1] Qin-Yi Tong,[1,2] Ulrich M. Gösele[1,3]

[1] Wafer bonding Laboratory, Department of Mechanical Engineering and Material Science, Duke University, Durham, NC 27708-0300
[2] Microelectronics Center, Research Triangle Institute, RTP, NC 27709
[3] Max-Plank Institute of Microstructure Physics, Weinberg 2, D-06120 Halle, Germany

Abstract

The thermal stress behavior of silicon/quartz bonded wafer pairs is examined. Sliding, debonding, and cracking are the observed mechanisms of relaxation. When the elastic energy due to the different thermal expansion coefficients of silicon and quartz exceeds the bonding energy, sliding will start and lead to a serrated curve on the curvature-versus-temperature graph. Finally, debonding will occur once the peeling stress exceeds the interface bonding strength. The debonded parts crack due to the overhang structure, and debonding-cracking processes continue during a further temperature increase. The stress behavior of the hydrogen-implantation induced layer splitting process (the so-called "Smart-Cut process") of silicon/quartz pairs is also monitored in a stress measurement setup. It is observed that Smart-Cut process is a sudden process in agreement with the observations reported in the literature.

Introduction

The fabrication of thin-film-transistor devices on a transparent substrate is key to active-matrix-liquid-crystal-display (AMLCD) technology. It has been suggested that the use of single crystalline-Si TFTs as pixel transistors will bring several benefits, including higher aperture ratio (for a brighter screen), lower power consumption, faster response time, and the possibility of substantial cost reduction. Moreover, in view of the required high processing temperature, quartz is a superior transparent substrate to glass [1]. Based on these factors, silicon on quartz fabricated by wafer bonding technology is a likely substrate candidate for AMLCDs.

"Direct wafer bonding" technology has received increasing attention since the middle of 1980s. With this technology, similar or dissimilar materials with clean, flat, smooth, and reactive surfaces can be brought into contact and form strong bonds without the use of an adhesive or of an electric field. Following the bonding process, subsequent annealing is commonly used to increase the bonding energy for a thinning process. Currently, several thinning techniques are suitable in connection with wafer bonding. Hydrogen implantation induced layer splitting [2] is an attractive approach because of the uniformity of the transferred layer thickness (also ± 10nm), and the possibility to reuse (after some light polishing) the wafer from which the transferred layers has been split. This thinning technique also requires a step involving elevated temperatures. In the case of dissimilar materials bonding such as silicon on quartz, a thermal mismatch is unavoidable during annealing because of the different thermal expansion coefficients of the two wafers involved. This thermal stress may lead to sliding, debonding, plastic deformation, thermal misfit dislocations, or even cracking, and directly influences the yield. Therefore, the present study of the silicon/quartz system was undertaken to follow the development of thermal strains and various relaxation phenomena during the annealing and layer splitting process.

Process Description

4" diameter and 400-μm-thick silicon wafers of (100) orientation were bonded to 4" 540-μm-thick quartz wafers at room temperature in a microcleanroom setup. Before bonding, the silicon wafers were cleaned in RCA-1 solution, and the quartz wafers were treated in ammonium hydroxide solution to increase the bonding energy [3]. Then, the wafer pairs were annealed at 150 °C to increase the bonding energy for further handling. To monitor the curvature change of the bonded wafer pairs during the annealing process, the samples were positioned in Tencor stress measurement tool which was equipped with a heating setup. The temperature was then increased up to 325 °C with 1 °C/min increment.

Results and Discussion

To study the thermal strain as well as relaxation phenomena in situ, the bonded wafer were subjected to temperature cycles from room temperature up to 325 °C. The curvature versus temperature during a measurement cycle is shown in Fig. 1. The initial smooth increase of curvature becomes serrated above 160 °C. Since the wafer pairs showed no visible damage during this period, sliding is likely to be responsible for the serrated curve. In fact, the given $1/R=0.44$ m^{-1} yields an elastic energy E=4.15 J/m^2, which exceeds the interface energy measured as 3.2 J/m^2 at this temperature [4]. Therefore, it could be expected that relaxation by interface sliding process would start, as also suggested by Cha [5]. The serrated curve should in fact have started already at a lower temperature for which an equal value of elastic energy and the measured interface energy was reached. Several possibilities may account for this delayed sliding. The silicon/quartz pair for bonding energy measurements was not processed in ammonium solution whereas this was done for the stress measurement samples. Therefore, the bonding energy may be smaller for the reference sample than for the silicon/quartz pair for stress measurement. The other possibility is that sliding may involve a nucleation phenomenon and therefore actually may start only at temperature higher than that at which the elastic energy equals the bonding energy.

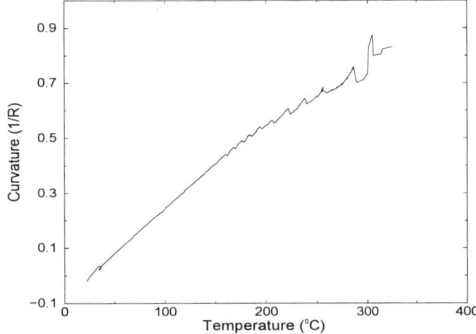

Fig. 1. Curvature of a 400μm Si/525μm quartz wafer pairs measured during a heating cycle.

Above 250 °C, the silicon/quartz wafer pair started debonding from the rim in a random fashion. After debonding to a certain degree, the debonded silicon part would crack, and the debonding-cracking process would continue up to the end of the annealing cycle. A photo of the fractured bonded pair is shown in Fig. 2. This cracking behavior is definitely different from that of silicon on sapphire (SOS) [6]. For a bonded pair of silicon on sapphire, the silicon wafer cracked into small fragments along the {100}-type planes that mostly remained bonded. This difference can be explained by a stress analysis of these two different bonding combinations. For the silicon on quartz bonded pair, the silicon has a larger thermal expansion coefficient (3.95×10^{-6}) than quartz (0.55×10^{-6}). Therefore, the silicon is under compressive stress at elevated temperatures. In contrast, the silicon wafer in an SOS pair is under tensile stress at elevated temperatures because of the higher thermal expansion coefficient of sapphire (8.4×10^{-6}). Generally speaking, brittle materials such as silicon can withstand higher compressive than tensile stress values. Therefore, for the silicon/quartz bonded pairs, the peeling stress, which is acting perpendicular to the bonding interface, played a more important role in the cracking phenomena, while the normal stress, which is the spatial distribution of biaxial normal stresses parallel to the bonding interface, might account for the crystal-orientation-type cracks for SOS pairs.

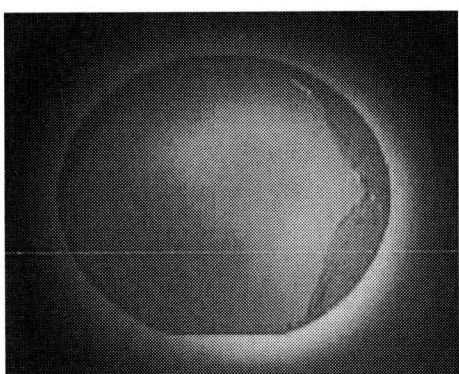

Fig. 2. The fractured Si/quartz bonding pairs after annealing cycle to 325°C.

Since the debonding-cracking phenomena for the silicon-quartz pair occurred above 250°C, the peeling stresses distribution at 300 °C was calculated based on the equations derived by Mirman et al. [7], and is shown in Fig. 3. A tensile stress as high as 13 MPa is presented at the edge of the sample. According to Berthold et al. [8], Si/Si bonding can yield a bond strength of 7-15 MPa at 300 °C. The bonding strength of a silicon/quartz bonded pair was slightly smaller than that of Si/Si bonded pairs. Therefore, it is quite reasonable to assume that the value of the peeling stress was high enough to overcome the bonding strength and cause delamination. However, the secondary peak in Fig. 3 indicates that a compressive stress probably limits a fast growth of the debonded area. Just as observed in the experiment, growth of the delaminated area should be a slow process. The debonded part formed a so-called overhang structure. Based on finite element models of the bonded wafer, the overhang structure induces a high tensile stress concentration right at the remaining bonded wafer edge. The stress drops immediately to compressive stress along the bonding interface toward the wafer center [9]. This stress distribution can easily result in cracking, as observed in the experiment.

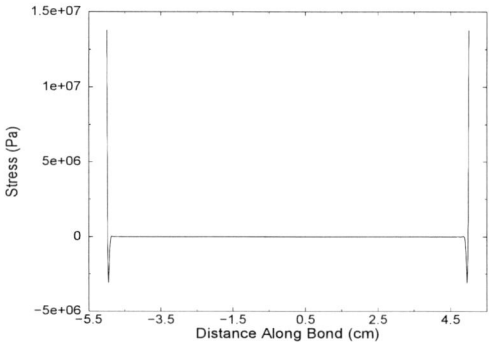

Fig. 3. Peeling stress distribution of Si/quartz wafer pairs at 300°C.

Stress Variation in Smart Cut[®] Process

Based on the observation of the above silicon/quartz bonding pair, the stress variation of a hydrogen-implantation-induced layer splitting process was undertaken with two pairs of 327-μm-thick silicon and 540-μm-thick quartz. In the silicon wafer of one pair, boron and hydrogen were co-implanted at relatively moderate temperatures [10]. The other pair involved a silicon wafer without implantation as a reference. In Fig. 4, it can be observed that the two wafer pairs have similar stress variations as a function of temperatures, whether the silicon wafer was implanted or not. Both of them had the same curvature change starting from room temperature and had sliding occurring at about 150°C. For the hydrogen-implanted sample, the significant drop at 230 °C from curvature 0.7 m^{-1} down to 0.05 m^{-1}, which equals the initial wafer curvature at room temperature, indicates that the layer splitting occurred in a short period of less than 2 minutes which was the measurement interval. This result suggests that the splitting is a sudden pop-up step rather than a gradual growth, supporting the results reported by Huang *et al.* [11].

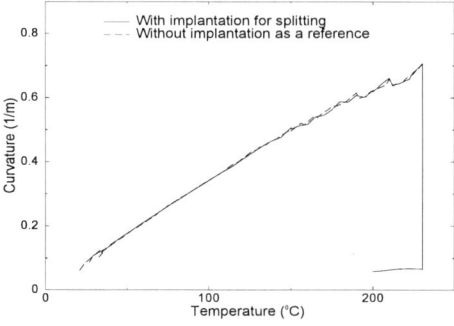

Fig. 4. The curvature of 327μm Si/525μm quartz wafer pairs with and without implantation measured during a heating cycle.

Conclusion

In conclusion, the thermal stress in bonded silicon/quartz wafer pairs can be relaxed via sliding, debonding, and cracking. These stress relaxation mechanisms are closely related to the bonding energy of the bonded wafer pair. Careful design of the heating process allows optimizing the product quality and yield.

Acknowledgment

This work was prompted by stimulating discussions with Dr. G. Kästner and Dr. P. Kopperschmidt. This work was partly supported by the German Federal Department of Science and Technology.

References

1. J. S. Im, R. S. Sposili, MRS Bulletin, 39 (March 1996).
2. M. Bruel, B. Aspar, C. Maleville, H. Moriceau, T. Poumeyrol, Proceedings of the 2nd Symposium on Advanced Science and Technology of Silicon Materials, 214, (1996).
3. Y.-L. Chao, Master thesis, Duke University, N.C., 1998.
4. T.-H. Lee, Ph. D. thesis, Duke University, N.C., 1998.
5. G. Cha, Ph. D. thesis, Duke University, N.C., 1994.
6. P. Kopperschmidt, G. Kästner, D. Hesse, N. D. Zakharov, and U. Gösele, Appl. Phys. Lett., **70**, (22), 2972 (1997).
7. B. A. Mirman, S. Knecht, IEEE Trans. On Comp. Hybrids, and Manuf. Tech., **13**, 914 (1990).
8. A. Berthold, M. J. Vellekoop, Sensors and Actuators, **A60**, 208 (1997).
9. M. E. Grupen-Shemansky, G. W. Hawkins, and H. M. Liaw, Proc. 1st Int. Sym. Semicond. Wafer Bonding: Science, Technol. and Appl., The Electrochem. Soc. **PV92-7**, 132 (1992).
10. Q.-Y. Tong, R. Sholz, U. Gösele, T.-H. Lee, L.-J. Huang, Y.-L. Chao, T. Y. Tan, Appl. Phys. Lett., **72**, (1), 49 (1998).
11. L.-J. Huang, Q.-Y. Tong, Y.-L. Chao, T.-H. Lee, T. Martini, and U. Gösele, Appl. Phys. Lett., **74**, (7), 982 (1999).

Mat. Res. Soc. Symp. Proc. Vol. 681E © 2001 Materials Research Society

Anodic Bond Quality and impact on Pressure Sensor Long-Term Stability

Henry Allen[1], Kamrul Ramzan[1], Jim Knutti[1], Carl Ross[2], Tim Milliman[3], Jeff Frye[2]
[1]Silicon Microstructure Incorporated, 46583 Fremont Blvd. Fremont, CA 94538
[2]Motorola AIEG, 4000 Commercial Ave., Northbrook, Il 60062
[3]Motorola AIEG, 611 Jamison Road, Elma, NY 14059

Abstract: Silicon pressure sensors have historically been fabricating by bonding a glass wafer to a micro-machined silicon wafer. The sensor may be sealed as an absolute pressure sensor by using planar glass and can then be used for detection of barometric pressure changes.

It has generally been assumed that as long as the glass and silicon are reasonable clean, then the silicon-glass seal is good and the part becomes a reliable, stable sensor. This paper addresses a low-level drift that was identified in such an absolute pressure sensor. A Zero drift in the range of 0.1% FS was detectable under humidity stresses. The stress always caused drift in the same direction, indicating an effective increased pressure in the sealed cavity.

The impact of various cleaning processes in reducing drift are reported. The improved process assure reliable product for applications such as automotive and altimeter applications.

INTRODUCTION

A large number of pressure sensors are still formed using conventional anodic glass-silicon bonding. This process simply requires applying a voltage of about 600 to 1000 volts across a glass-silicon sandwich at an elevated temperature of a 300 to 400 C. These pressure sensors are found in a variety of applications from very low pressure systems (<10 mBar) to very high pressure (>300 Bar). One particular pressure range is in atmospheric pressure measurements; applications include barometric sensing, altimeters, and fuel mixture correction in vehicles based on atmospheric pressure. This structure generally uses a 1 Bar pressure sensor and a flat glass wafer is bonded to the back of the etched silicon wafer as shown in Figure 1.

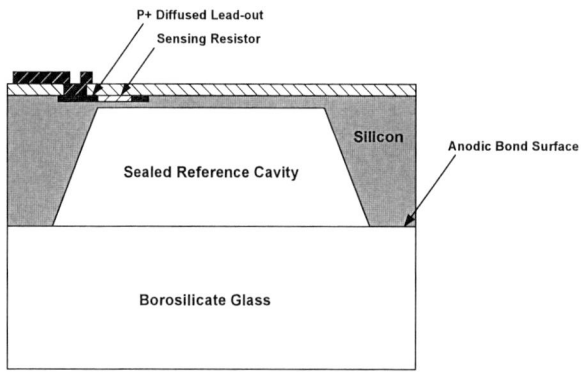

Figure 1. Cross-section of an absolute pressure sensor

The particular design issue addressed in this paper relates to a drift observed during the assembly of an automotive pressure sensor. Typically, the sensor gives out a 60 mV full-scale output for 0 to 15 PSIA. The sensor is generally protected from the automotive environment by a silicone-based gel that coats the die. See Figure 2 as an example. The general requirement for acceptable stability is about 100 microvolts of drift after various stresses. During stress testing for humidity, it was observed that typically 1 to 3% of parts from some lots drifted enough to be detected (typically 100 to 500 microvolts drift). These parts, when the protective gel was removed, showed essentially no drift. Thus, an apparent conclusion was that the gel was contributing to the drift. What was also determined was that stressing the part once caused the largest shift but letting the module rest tended to restore some, but not all of the offset back to its original target. Subsequent stressing would continue to drive the sensor output downward.

Figure 2a. Sensor coated with gel   **Figure 2b.** Packaged Sensor System

As a measure of stability of the product, modules are randomly sampled at the end of the production line to assure conformance both to parametric and short-term errors. One of those short-term tests done as an audit of product quality is humidity drift over a couple of day stress.

A substantial amount of effort was directed at isolating the true cause of the drift. The problem was made more difficult to trace down due to a special metal process used in this sensor that could, due to surface leakage, contribute to the problem. One of the observed phenomena was that the sensors always drifted in one direction. This was such that it might look like there was a slight (less than 0.075 PSI increase in pressure in the cavity). To compound isolating the problem, however, the resistor structure was asymmetric so a drift could also be related to the circuit topology and not related to a mechanical flaw.

PROCESS DESCRIPTION

The process involves fabricating a piezoresistive pressure sensor in a 2.25 mm die size. The process is unique because the environment in which the die is used has highly corrosive vapors from engine fuel components and their derivatives. Normal aluminum metalization did not last in this specific application so a gold metalization was utilized. A PtSi contact was used to make good ohmic contact to the silicon for the device and then a Ti/TiW/Au metalization was deposited. Because of the adhesion characterisitics of gold, the Ti/TiW/Au metal was deposited on the top surface of the silicon nitride. This led to the possibility that surface leakage due to incomplete cleaning or residual PtSi or Ti could cause surface leakage. A photomicrograph of the sensor is shown in Figure 3. The large "L" shape metal in the upper-right hand corner serves as a field-shield for an offset resistor that causes an intentional imbalance in the bridge and thus makes the sensing network asymmetric.

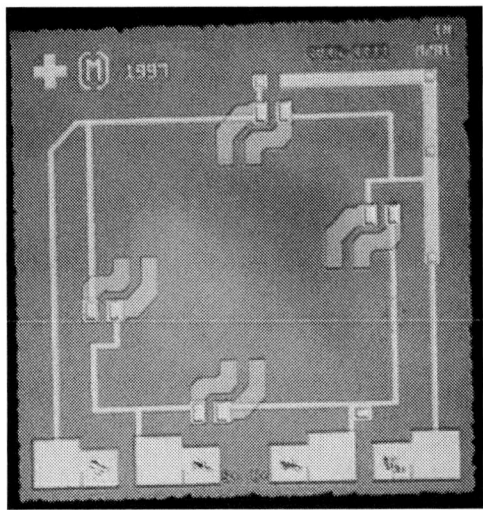

Figure 3. Top View of Pressure Sensor Die

A cross-section of the top surface of the structure is shown in Figure 4. It must be remembered that tracing down the drift to the anodic bond quality was not an obvious issue. In the various ratings, most effort was actually placed on other sources of drift.

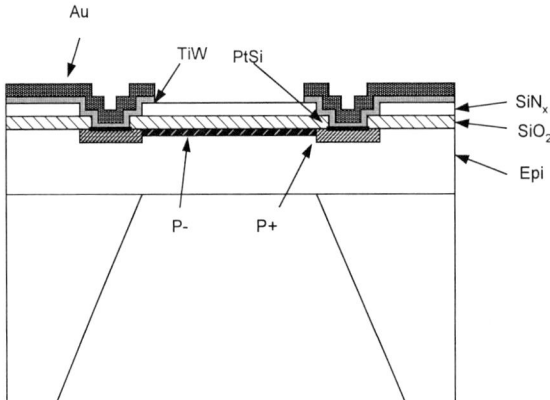

Figure 4. Cross-section of top surface of sensor.

EXPERIMENTAL CAUSE ISOLATION

A series of tests were performed to isolate the cause of the drift. Based on an observation that the gel coating was a high contributor to the cause of the drift, most of the tests involved surface tests initially. In fact, the anodic bond was rated very low in the potential risks. The most likely sources of suspected drift were contact resistance, surface leakage due to the PtSi process, and charge storage either in the silicone-gel or in the interface between the SiN_x and the SiO_2. The anodic bond was discounted because there was no apparent reason
- why gel should make the bond worse;
- why, if it were a weak bond, the drift should always be downward (more pressure in cavity); if it were a weak bond, then altering stress should result in offsets changing both up and down; or
- why it would be a random failure both die to die and wafer to wafer.

To isolate some problems, an IS/IS_Not Table (Table 1) was created to summarize what was known. The last entry (most likely associated with wafers with poor anodic bonds) was added very late in the analysis process.

Table 1 - IS/IS NOT
Modules fail when exposed to humidity

IS	IS NOT
On most lots of Noble Metal (no two sequential lots pass ELMA tests)	On Motorola Aluminum Material
On some wafers worse than others within same lot	Specific to Foundry Lot starts
On some lots worse than others	Specific to ECE lots
Larger change in some cases than others	Specific to Metal deposition lots
Always downward shift	One-step Contact Process
In some cases apparently associated with platinum or other contaminant residue on surface	Eliminated with longer dips in HF to remove Platinum residues (15 sec vs 60 sec test)
More likely on wafers with missing numbers (early observation)	Specific to metal pattern with larger or smaller P+ contact spacing separation
Most readily apparent when gelled parts are subjected to humidity	Specific to amount of haze on wafer
More likely on the edge of some wafers in crescent near major flat	Specific to oxide thinning amount (matrix lot of 1500 to 3500 A material)
Less likely on wafers with Engineering supervision	Eliminated with Saw-thru Silicon
Possibly on aluminum wafers (too small a sample, too long ago to tell; Aluminum parts were not as stable as control parts but comparable to Noble Metal parts)	Eliminated with Silicon Electrode (and higher bond current) with anodic bond
More likely associated with wafers with poor anodic bond	

Ultimately, the one primarily variable that was successfully found to turn-on and turn-off the problem was the anodic bond.

ANODIC BOND CLEANS AND THE IMPACT ON STABILITY

The original process was to protect the front surface of the silicon wafer with yellow tape while the backside had an HF Strip. The HF successfully stripped any oxide and nitride at the same time. After that, the wafers were cleaned in acetone to remove the tape followed by photo-resist stripper. An alternate of replacing the photo-resist stripper with IPA was also tried. The clean that was adopted was a sulfuric-peroxide clean after the acetone clean, followed by a 100:1 HF dip and a rinse and then spin-dry. The aggressive nature of the clean was made possible due to the TiW/Au metalization used on this device.

Both of the solvent based cleans proved to be ineffective in eliminating the drift. Of the 3 approaches, only changing to the sulfuric-peroxide clean gave consistent no-failure results at the humidity test. Further, based on tests in assembly, the drift effects could be turned off and on by switching from one clean to the other clean and back again. This was demonstrated three times. Each time the wafers were cleaned with the old, solvent-based clean, the likelihood of humidity failure increased and each time that the sulfuric-periodic clean was implemented, humidity failure rates went to zero.

CONCLUSIONS

Drift in the output of gelled pressure sensors, subjected to humidity stresses, has been used as an indicator of long-term drift. A systematic study of both anodic bond conditions and cleanliness of the die has lead to the conclusion that the surface preparation of the anodic bond interface is much more critical than previously thought.

What is thought to explain the phenomena is the following: the gel tended to collect and retain water-vapor. The water-vapor at elevated temperature diffuse through the weak anodic bond. The water-vapor molecules, even with baking-out the device, do not tend to escape the cavity and tend to maintain a fairly consistent partial pressure with the outside.

The actual mechanism of what caused the degraded anodic bond appears to be residual organics that are not properly cleaned during the final anodic bond clean, using the weaker solvent cleans. Residual organics could be locally present and would thereby address why population and density of failures could vary lot-to-lot and wafer-to-wafer.

Using an aggressive clean such as sulfuric-peroxide, followed by a 100:1 HF dip, and then a rinse and dry has been shown to be superior to solvent-based cleans for the clean immediately prior to anodic bond.

Since the new clean has been implemented, approximately 2.5 million modules have been built without any humidity drifts failures at out-going audit.

Mat. Res. Soc. Symp. Proc. Vol. 681E © 2001 Materials Research Society

A Novel Method of Fabricating SiC-On-Insulator Substrates for Use in MEMS

Hung-I Kuo, Christian Zorman, and Mehran Mehregany
Department of Electrical Engineering and Computer Science
Case Western Reserve University
Cleveland, OH 44106, USA

ABSTRACT

This paper reports on a novel, bonding-free method to fabricate silicon carbide-on-insulator (SiCOI) substrates. The process bypasses wafer bonding by using a high deposition rate polysilicon process in conjunction with wet chemical etching to produce wafer-thick polysilicon layers that serve as substrates for the SiCOI structures. Because wafer bonding is not used, insulators of various material types and thickness can be used. Using this method, transfer percentages over 99% are readily achievable. Various applications could benefit from this technology, including high temperature SiC-based microelectromechanical systems (MEMS) and SiC electronic devices.

INTRODUCTION

Silicon carbide (SiC) is well known as a wide-band-gap semiconductor with outstanding physical and chemical properties. Compared to silicon (Si), SiC exhibits a larger bandgap, a higher breakdown voltage, a higher thermal conductivity and higher saturation velocity. SiC is also considered to be leading candidate for high temperature microsensor and micro actuator systems, since the 3C-SiC polytype can be epitaxially grown on Si substrates and polycrystalline 3C-SiC (poly-SiC) films can be deposited on SiO_2 and polycrystalline silicon (polysilicon) sacrificial substrate layers [1]. 3C-SiC growth on (100) Si substrates enables the fabrication of bulk micromachined 3C-SiC structures, such as membranes and cantilevers, by selective removal of the Si substrate in various anisotropic etchants. Poly-SiC, when used in conjunction with SiO_2 and polysilicon, enables the fabrication of surface micromachined devices such as electrostatically driven lateral resonant structures and micromotors. Such devices are difficult to fabricate in single crystalline 3C-SiC films since although the Si substrate could be used as a sacrificial layer, it is not sufficiently insulating to support electrostatic actuation.

Silicon-on-insulator (SOI) substrates have great potential to enhance the performance of Si microelectronics, especially in the areas of power consumption, operating frequency, and operating temperature due to a significant reduction in leakage currents. Electronic devices fabricated from 3C-SiC films grown on SOI substrates should also have much lower leakage currents, a significant problem that has limited the use of 3C-SiC as an electronic material. Such a substrate would also facilitate the fabrication 3C-SiC MEMS devices, such as the electrostatically actuated devices described previously.

Currently, there are four methods that have been used to produce SiCOI substrates. The most direct method is epitaxial growth of 3C-SiC on various commercially available SOI wafers. These substrates include silicon-on-sapphire (SOS) [2], SIMOX® [3], and UNIBOND®(Smart-Cut®) [4]. The second method combines 3C-SiC growth with Si wafer bonding and etch-back [5] A third method involves epitaxial lateral overgrowth (ELO) of 3C-SiC films on Si substrates masked with SiO_2 and Si_3N_4 films [6]. The fourth method utilizes the Smart-Cut® process performed on 6H-SiC wafers [7]. Each of these methods has certain limitations with respect to yield, cost, material quality or other factors. In this paper, we report the development of a novel fabrication method to produce high-yield, large area SiCOI substrates using deposited polysilicon as the substrate material.

FABRICATION PROCEDURE

Figure 1 shows schematic cross-sections of the key steps in the fabrication process. The process begins with the acquisition of 100-mm diameter, 700 micron-thick silicon (100) wafers. Each wafer is cleaned using a DI water rinse to remove any particles and to otherwise prepare the wafers for SiC growth. A 3C-SiC film is heteroepitaxially grown on each wafer in a rf-induction-heated, cold wall reactor with a SiC-coated, graphite susceptor sized to grow films on two 100 mm-diameter Si wafers. 3C-SiC films are grown using a conventional, carbonization-based two step atmospheric pressure chemical vapor deposition process detailed elsewhere [8]. The process uses hydrogen as a carrier gas, propane as a carbon source and silane as the silicon source. Phosphine is used as an n-type dopant when required. Before film growth, the wafers are heated to 1000°C for 5 min in a hydrogen atmosphere in order to reduce the native oxide on the Si surface. The susceptor temperature is then reduced to below 500°C, at which point propane is added to the hydrogen flow and the susceptor temperature is raised to 1280°C. The gas flows and temperatures are held constant for 90 s, during which a thin 3C-SiC layer is formed by the carbonization of the Si surface. Film growth is initiated by adding silane to the gas flow and properly adjusting the propane flow rate. For this study, the 3C-SiC film thickness ranged between 0.5 and 2.0 microns.

Following film growth, the surfaces of the 3C-SiC films are mechanically polished to reduce surface roughness. The mechanical polishing recipe is designed to polish the surface of 3C-SiC films with a minimum of material removal. The polishing slurry consisted of a mixture of 6000 grit SiC powder and DI water. Ten minutes of polishing is sufficient to produce the desire surface finish. After polishing, the wafers are cleaned and readied for the deposition of an insulating thin film. In this study, we selected thermally grown SiO_2 and LPCVD Si_3N_4 thin films to serve as the insulating layers. Thermal SiO_2 layers are prepared by the deposition and complete thermal oxidation of polysilicon thin films. In this manner, a thermal oxide thickness in excess of 1.5 microns could be grown in a reasonable amount of time on the 3C-SiC surfaces, a process not feasible from a time and material point of view if direct thermal oxidation of 3C-SiC were used. The Si_3N_4 thin films were prepared using a conventional LPCVD process used to deposit stoichiometric Si_3N_4. For this reason, the maximum thickness of the nitride layer was about 2000Å, since high stresses in stiochiometric Si_3N_4 thin films lead to severe cracking in thick nitride films.

At this point, the wafers are ready for the deposition of the polysilicon substrate. Polysilicon "films" with a thickness on the order of 600 µm are deposited on the wafers. The polysilicon deposition process uses a silane-based recipe to deposit thick films at high deposition rates. The net effect of the deposition process is a polysilicon layer that is nearly stress balanced with the single crystal wafer, in part because the thermal mismatch is negligible, but also because the thickness of each silicon layer is roughly equivalent. The process essentially sandwiches the 3C-SiC and insulating thin films between two very thick Si layers. After this step, the wafer is ready for the 3C-SiC transfer etch. First, a protective Si_3N_4 or SiO_2 coating is deposited on the backside of the polysilicon substrate. The entire sample is then immersed in a Si wet etchant, such as a KOH/water solution, which etches the unmasked (100) Si wafer and stops on the 3C-SiC film, resulting in the creation of a 3C-SiC-on-insulator substrate. If needed, additional 3C-SiC can easily be grown on the 3C-SiC surface by a homoepitaxial process [9].

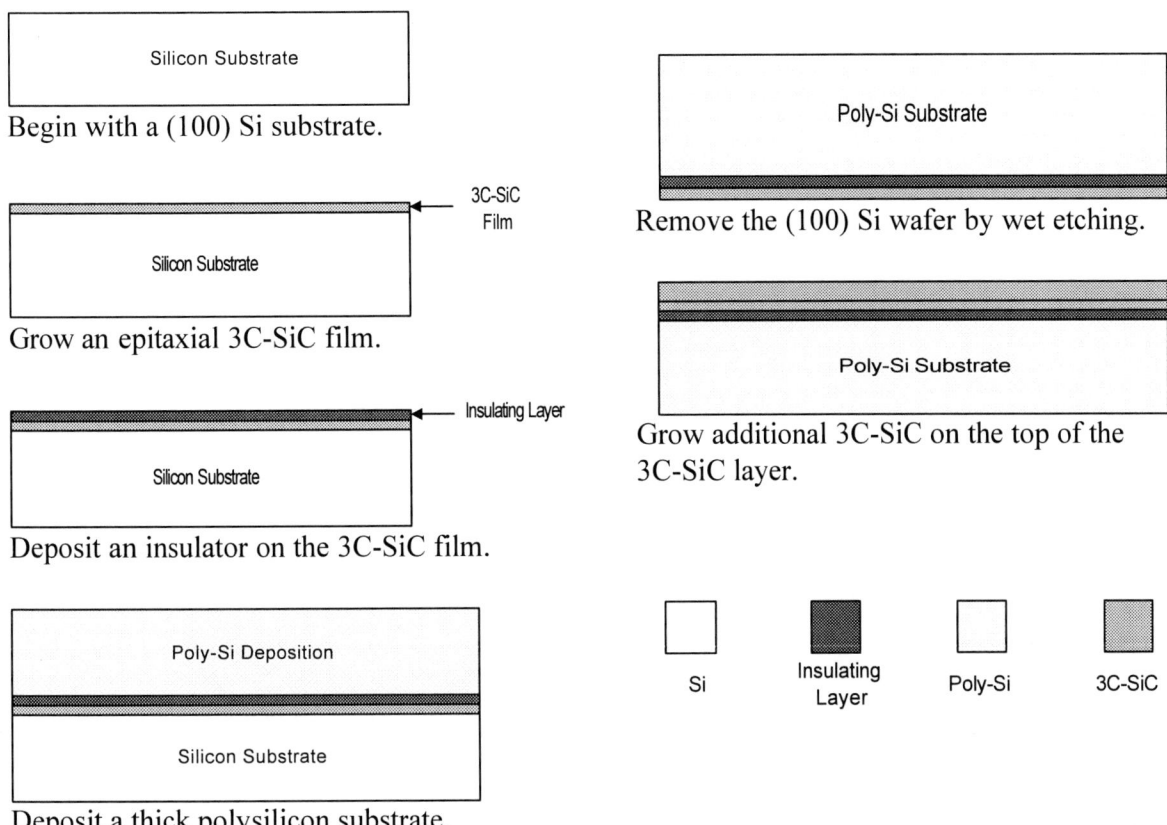

Figure 1. Cross-sectional schematics of the SiCOI-on-polysilicon fabrication process.

RESULTS

Figure 2 is an optical photograph of a SiCOI wafer formed on a polysilicon substrate, with the 3C-SiC surface facing outward. The 3C-SiC film is continuous across the entire substrate, indicating that the transfer yield is nearly 100%. This is not entirely surprising since the deposition of the polysilicon substrate is conformal to the 3C-SiC stress-induced curvature of the (100) Si wafer. 3C-SiC films grown on (100) Si substrates tend have high tensile stresses, generally in the range of 100 MPa to 300 MPa. Such high tensile stresses in combination with a single-sided 3C-SiC growth process, produces significant bowing (> 10 microns) of large area wafers. Wafer bow is problematic for 3C-SiC based wafer bonding processes, since tensile stress produces wafers with concave profiles, thus increasing the likelihood that trapped air pockets will form at the bonding interface. Trapped air pockets are not expected in our process since the transfer interface is formed by the growth of the final substrate on the original substrate, not by the joining of two separate substrates.

Figure 2. Optical photograph of a 100 mm-dia. SiCOI wafer formed on a polysilicon substrate.

Cross-sectional transmission electron microscopy (TEM) was performed on one of the substrates to examine the crystalline properties of the 3C-SiC film and the interface properties of the SiCOI multilayer substrate. The TEM micrograph in Fig. 3 shows a 3C-SiC film with a SiO_2 insulating layer on the polysilicon substrate. The TEM shows a clear contrast between each of the layers. The 3C-SiC film shows a pattern of defects characteristic of 3C-SiC growth on Si wafers, while the SiO_2 film appears as a well-formed amorphous layer. The polysilicon substrate has a randomly-oriented grain structure characteristic of polysilicon grown at high temperatures, which is expected since high deposition rates generally require relatively high processing temperatures. As expected, sharp and featureless interfaces are formed between the each of the three layers, with no cavities or other defects sometimes seen in wafer bonded samples.

Figure 3. Cross sectional TEM micrograph of a SiCOI-on-polysilicon substrate.

In order to demonstrate the durability and versatility of the SiCOI substrate, a surface micromachining process originally used to fabricate 3C-SiC lateral resonators on wafer bonded substrates [10] was performed on the polysilicon substrates. A cross-sectional schematic of the fabrication process is shown in Fig. 4. The process uses a 3C-SiC film as the structural layer, a SiO_2 film as a sacrificial and electrical isolation layer, and a polysilicon substrate for mechanical support. The 3C-SiC film was patterned using conventional photolithography and a reactive ion etching process based on a $CHF_3/O_2/He$ plasma chemistry. Aluminum was employed as a hard mask. The resonators were released using a timed HF etch. A SEM micrograph of a released resonator is shown in Fig. 5. This resonator has a resonant frequency of 18 kHz, which is expected from the material properties and geometry of the structure. Close examination of the micrograph reveals that the film has a residual stress gradient, as evidenced by the fact that the folded beam structures are not co-planar with the shuttle and anchor pads.

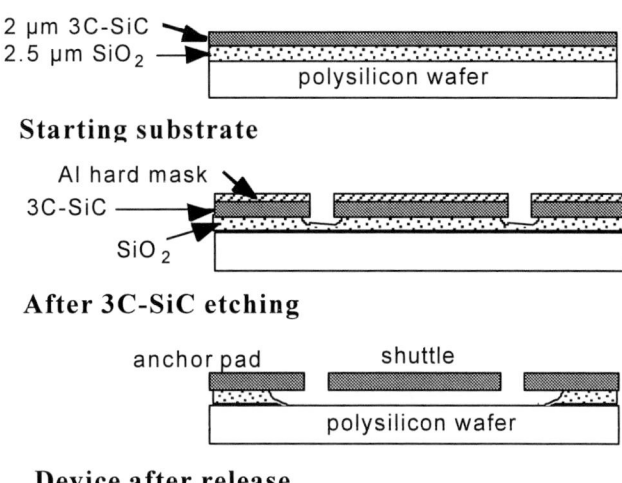

Figure 4. Cross-sectional schematics of a surface micromachining process on SiCOI wafers.

Figure 5. SEM of a 3C-SiC lateral resonator fabricated on a SiCOI-on-polysilicon substrate.

CONCLUSIONS

This paper reports on a novel, ion-implantation and bonding-free method to fabricate SiCOI substrates. The process combines 3C-SiC epitaxial growth with a high deposition rate, polysilicon process and wet chemical etching to produce SiCOI structures on wafer-thick polysilicon substrates. With this unique method, insulators of various material types (SiO$_2$ and Si$_3$N$_4$) and thickness can be used. A transfer percentage of over 99% can be achieved because the process is insensitive to wafer bow. Various applications could benefit from this technology, including MEMS sensors (e.g. pressure sensors, chemical sensors), actuators (resonators), and SiC electronic devices.

ACKNOWLEDGEMENTS

The authors thank Dr. Juyong Chung and Mr. John Sears of the Department of the Materials Science and Engineering at Case Western Reserve University for assisting with the TEM and SEM analysis. This work was supported by DARPA under contract No.DABT63-98-1-0010.

REFERENCES

1. M. Mehregany, C.A. Zorman, S. Roy, A.J. Fleischman, C.H. Wu, and N. Rajan, *International Materials Review,* **45**, (2000), 85-107.
2. J.C. Pazik, G. Kelner, N. Bottka, J.A. Freitas Jr., *Material Science Engineering* B **11** (1992) 125.
3. J.M. Bluet, S.Contreras-Azema, J. Camassel, J.L. Robert, L. Di Cioccio, W. Reichert, R. Lossy, E. Obermeier, J. Stoemenos, *Materials Science and Engineering* B46 (1997) 152-155
4 J. Camassel, Journal of Vacuum Science & Technology B **16**, (1998), 1648-1654
5 K. Vinod,"3C-SiC on Silicon Oxidation, Wafer Bonding, and Schottky Diodes" Department of Electrical Engineering and Applied Physics Thesis (M.S.)--Case Western Reserve University, 1997.
6 C.H Wu, "Growth and Characterization of Silicon Carbide for MEMS Pressure Sensors" Department of Materials Science and Engineering, Thesis (Ph.D.)--Case Western Reserve University, 2001
7 L. DiCioccio, Y. LeTiec, F. Letertre, C. Jaussaud, M. Bruel, *Electronics Letters*, 32 (12) (1996) 1144-1445
8. C.A. Zorman, A.J. Fleischman, A.S. Dewa, M. Mehregany, C. Jacob, S. Nishino, and P. Pirouz, *Journal of Applied Physics*, **78**, (1995), 5136-5139.
9. K. N. Vinod, C.A. Zorman, A. Yasseen, and M. Mehregany, *Journal of Electronic Materials,* **27**, (1998), L17-L20.
10. S. Stefanescu, A.A. Yasseen, C.A. Zorman and M. Mehregany, *Proceedings of the 10th International Conference on Solid State Sensors and Actuators,* Sendai, Japan, June 7-10, 1999, pp. 194-197.

Wafer Bonding between Magnetic Garnet and Lithium Niobate
for Semi-Leaky Isolator

Hideki Yokoi, Tetsuya Mizumoto and Masafumi Shimizu
Department of Electrical and Electronic Engineering, Tokyo Institute of Technology,
2-12-1 Ookayama, Meguro-ku, Tokyo, 152-8552 JAPAN,
yokoi@wave.pe.titech.ac.jp

ABSTRACT

A semi-leaky isolator composed of a magneto-optic / anisotropic waveguide was fabricated by wafer bonding technique. Characteristics of the semi-leaky isolator were investigated. The magneto-optic / anisotropic waveguide exhibited the semi-leaky characteristics. The isolation ratio of 7.2 dB/cm was obtained at a wavelength of 1.55 μm.

INTRODUCTION

In optical communication systems, an optical isolator is indispensable in protecting optical active devices from unwanted reflected light. In the near infrared region, magnetic garnet crystals are necessary to construct the optical isolator. A waveguide-type optical isolator is desired for photonic integrated circuits. Various kinds of the waveguide-type optical isolator have been investigated by many researchers.

A semi-leaky isolator was proposed by Yamamoto et al. in 1976 [1]. The semi-leaky isolator utilized unidirectional mode coupling in a magneto-optic / anisotropic waveguide. The semi-leaky isolator was attractive because of its compact mono-section structure and easy control of magnetization. However, the operation of the device was not confirmed due to poor optical contact between the magneto-optic guiding layer and the anisotropic cladding layer [1]. Kirsch et al. demonstrated the operation of the semi-leaky isolator by using a selenium intermediary layer between the magneto-optic layer and the anisotropic layer [2].

The authors have investigated wafer bonding between magneto-optic crystals and III-V compound semiconductors with the aim of integrating a semiconductor laser diode and an optical isolator [3,4]. It was considered that the optical contact between the magneto-optic crystal and the anisotropic crystal could be accomplished by this technique.

In this paper, we report on a semi-leaky isolator fabricated by wafer bonding technique. Wafer bonding between the magneto-optic crystal and the anisotropic crystal was studied. Characteristics of the semi-leaky isolator fabricated by wafer bonding were investigated.

DEVICE STRUCTURE

Figure 1 shows the basic geometry of the semi-leaky isolator. Cerium-substituted yttrium iron garnet (Ce:YIG) and lithium niobate (LiNbO$_3$) are assumed as a magneto-optic layer and an anisotropic layer, respectively. Faraday rotation coefficient of the Ce:YIG is approximately −4,500 deg/cm at a wavelength of 1.55 μm [5,6]. The optic axis of the lithium niobate is rotated in the film plane, which is done just by rotating the crystal. An external magnetic field is applied to the light propagation direction.

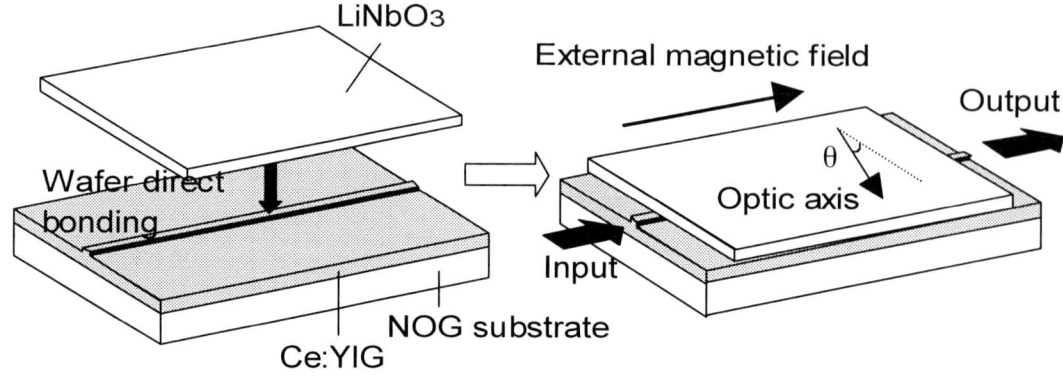

Figure 1. Basic geometry of semi-leaky isolator. The waveguide has a magneto-optic guiding layer and an anisotropic cladding layer.

Principle of the isolator operation is described from perturbation theory. In the absence of the perturbations, all dielectric tensors are assumed to be diagonal. An arrangement of the refractive index profile gives rise to the semi-leaky characteristics, that is, TE waves are guided and TM waves are leaky.

In reality, there are nonzero but small off-diagonal elements in the magneto-optic guiding layer and the anisotropic cladding layer by applying the external magnetic field and the optic axis rotation, respectively. An introduction of the off-diagonal elements gives rise to the coupling between TE guided and TM radiation modes with orthogonal polarization of the basic system. Since the electromagnetic field penetrates the LiNbO₃ cladding layer with the optic axis rotation, the anisotropic perturbation yields reciprocal coupling. The gyrotropic perturbation induced by the magneto-optic effect leads to nonreciprocal coupling. By adjusting the two elementary coupling coefficients, they are the same in magnitude but with opposite signs in the forward traveling waves. In this case, the coupling would occur in the backward traveling waves only so that this device provides a unidirectional mode coupling between TE guided and TM radiation modes.

The coupling coefficients can be controlled by adjusting the thickness of the magnetic garnet and the optic axis rotation of the LiNbO₃. Waveguide parameters were investigated for optimum condition to cancel the mode coupling for forward traveling waves. Figure 2 shows the offset angle of the crystal axis, which is a rotation angle from the crystal axis, depending on the Ce:YIG thickness. Since the mode coupling is canceled for forward traveling waves, the backward loss corresponds to the isolation ratio. When the Ce:YIG thickness is 0.9 μm, the coupling is canceled with the offset angle of 12.0°. Theoretical isolation ratio of more than 70 dB/cm is expected in the semi-leaky isolator with the 0.9-μm-thick guiding layer.

Characteristics of the semi-leaky isolator were investigated when the offset angle deviated during wafer bonding. Figure 3 shows the forward and backward loss of the semi-leaky isolator depending on the offset angle of the crystal axis. The thickness of the Ce:YIG was assumed to be 0.9 μm. The isolation ratio of the semi-leaky isolator is also shown in Fig. 3. Although a larger isolation ratio is achieved with a larger offset angle, the forward loss also becomes larger.

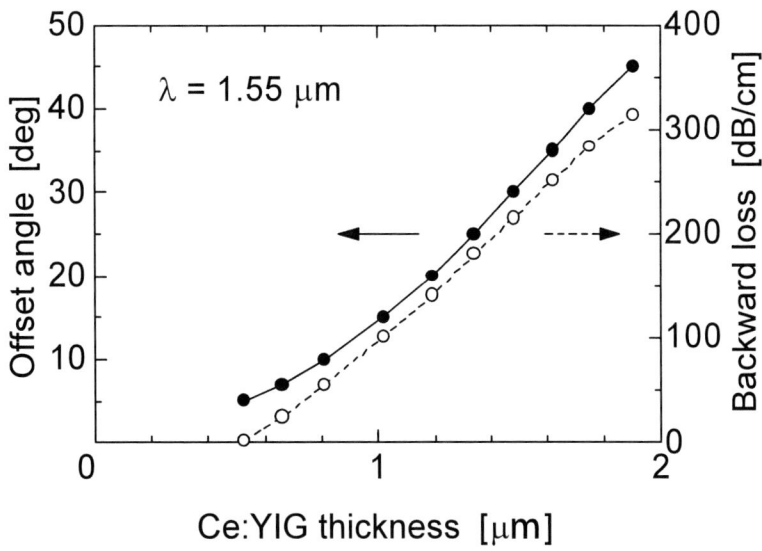

Figure 2. Offset angle of crystal axis rotation depending on Ce:YIG thickness for isolator operation. Backward loss of the semi-leaky isolator is also shown.

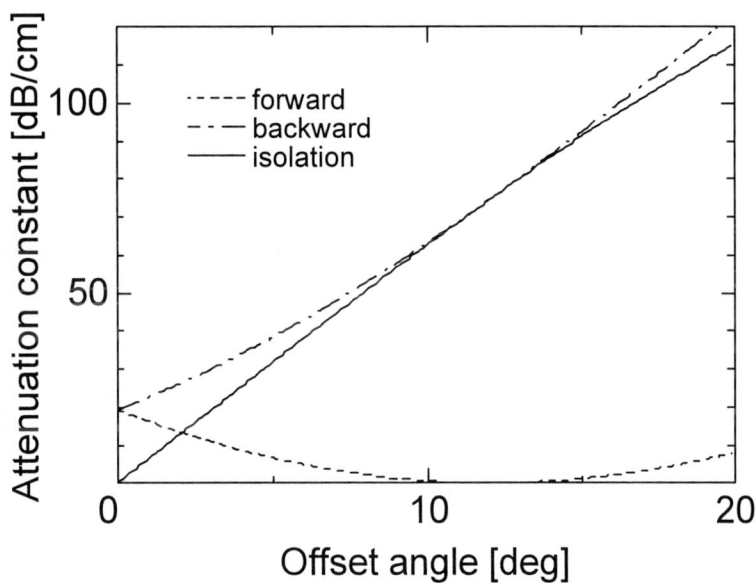

Figure 3. Attenuation constant depending on offset angle of crystal axis.

EXPERIMENTAL RESULTS

A Ce:YIG layer was grown on a garnet substrate by sputter epitaxy [7]. Wafer bonding between Ce:YIG and LiNbO$_3$ was studied. Ce:YIG was slightly etched by H$_3$PO$_4$ and LiNbO$_3$ was treated by NH$_4$OH:H$_2$O$_2$:H$_2$O to obtain hydrophilic surfaces. The two hydrophilic surfaces were connected at room temperature and loaded into an annealing furnace. Wafer bonding was successfully achieved with the heat treatment at 220ºC in H$_2$ ambient.

The semi-leaky isolator was constructed by wafer bonding technique. The rib waveguides were fabricated on the 0.9-μm-thick Ce:YIG by rf sputter etching with Ti mask. The rib height and the rib width were 0.1 μm and 3.0 μm, respectively. Wafer bonding between the Ce:YIG rib waveguide and LiNbO$_3$ was achieved with the heat treatment at 220ºC in H$_2$ ambient.

The propagation of light waves into the Ce:YIG / LiNbO$_3$ waveguide was studied by guided wave optics technique. The light waves from a semiconductor laser diode ($\lambda = 1.55$ μm) were made incident on the 5-mm-long waveguide using end-fire coupling via a single-mode fiber. The output from the waveguide was observed by a TV monitor through an optical microscope objective and IR camera. Figure 4 shows near field patterns of TE and TM modes at the output facet of the Ce:YIG / LiNbO$_3$ waveguide. It was ascertained that the Ce:YIG / LiNbO$_3$ waveguide exhibited the semi-leaky characteristics, that is, TE waves were guided and TM waves were leaky. The TM leaky modes indicate that the traveling light waves experience LiNbO$_3$ cladding layer.

The light waves of TE modes were launched into the semi-leaky isolator to measure the isolation ratio at 1.55 μm. An external magnetic field was applied along the light propagation direction to produce the magneto-optic coupling between TE guided and TM radiation modes. When the magnetic field was reversed, the output power from the output facet of the isolator varied. The reversal of the magnetic field was equivalent to the measurement of the counter-propagating waves. The isolation ratio was estimated to be approximately 7.2 dB/cm at 1.55 μm from the change of the output power.

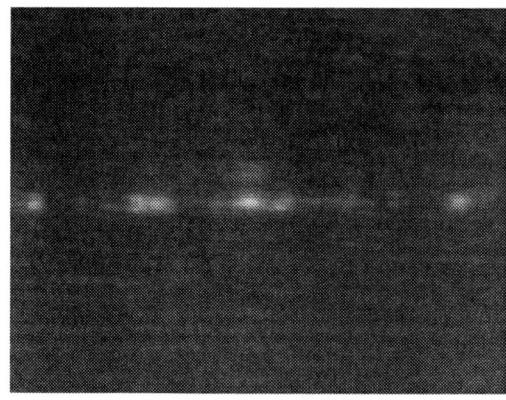

TE mode   TM mode

Figure 4. Near field patterns of TE and TM modes at the output facet of the Ce:YIG / LiNbO$_3$ waveguide. The waveguide exhibits the semi-leaky characteristics.

CONCLUSIONS

A semi-leaky isolator fabricated by wafer bonding technique was studied. The semi-leaky isolator utilized unidirectional mode coupling in a magneto-optic / anisotropic waveguide. The semi-leaky isolator was designed at a wavelength of 1.55 μm. Wafer bonding between a magnetic garnet and a lithium niobate was investigated to construct the semi-leaky isolator. The isolation ratio of approximately 7.2 dB/cm was achieved at 1.55 μm.

ACKNOWLEDGMENTS

The authors would like to acknowledge Dr. Y. Naito, Professor Emeritus, for his continuous encouragement. This work was partially supported by Exploratory Research on Novel Artificial Materials and Substances for Next-Generation Industries of the Research for the Future from Japan Society for the Promotion of Science (JSPS-RFTF97P00103).

REFERENCES

[1] S. Yamamoto, Y. Okamura and T. Makimoto, IEEE. J. Quantum Electron. **QE-12**, 764 (1976).
[2] S. Kirsch, W. A. Biolsi, S. L. Blank, P. K. Tien, R. J. Martin, P. M. Bridenbaugh and P. Grabbe, J. Appl. Phys. **52**, 3190 (1981).
[3] H. Yokoi, T. Mizumoto, K. Maru and Y. Naito, Electron. Lett. **31**, 1612 (1995).
[4] H. Yokoi, T. Mizumoto, M. Shimizu, T. Waniishi, N. Futakuchi, N. Kaida and Y. Nakano, Jpn. J. Appl. Phys. **38**, 4780 (1999).
[5] M. Gomi, S. Satoh and M. Abe, Jpn. J. Appl. Phys. **27**, L1536 (1988).
[6] T. Shintaku and T. Uno, J. Appl. Phys. **76**, 8155 (1994).
[7] T. Shintaku and T. Uno, Jpn. J. Appl. Phys. **35**, 4689 (1996).

Mat. Res. Soc. Symp. Proc. Vol. 681E © 2001 Materials Research Society

Integration of InGaN-based Optoelectronics with Dissimilar Substrates by Wafer Bonding and Laser Lift-off

William S. Wong, Michael Kneissl, David W. Treat, Mark Teepe, Naoko Miyashita, and Noble M. Johnson
XEROX Palo Alto Research Center, 3333 Coyote Hill Road, Palo Alto, CA 94304, USA

ABSTRACT

InGaN-based optoelectronics have been integrated with dissimilar substrate materials using a novel thin-film laser lift-off process. By employing the LLO process with wafer-bonding techniques, InGaN-based light emitting diodes (LEDs) have been integrated with Si substrates, forming vertically structured LEDs. The LLO process has also been employed to integrate InGaN-based laser diodes (LDs) with Cu and diamond substrates. Separation of InGaN-based thin-film devices from their typical sapphire growth substrates is accomplished using a pulsed excimer laser in the ultraviolet regime incident through the transparent substrate. Characterization of the LEDs and LDs before and after the sapphire substrate removal revealed no measurable degradation in device performance.

INTRODUCTION

Often the enhancement of integrated microsystems requires the integration of thin-film materials with disparate properties. Many examples may be found currently in the III-nitride optoelectronic field. For example, today's most advanced high-performance (In,Ga,Al)N laser diodes (LDs) possessing lifetimes greater than 10000 hours have been realized on sapphire substrates[1] although a major impediment to the development of III-nitride LDs still remains the efficient dissipation of heat generated from the device active area. The high thermal resistance of the sapphire substrate and the high diode current densities combine to degrade the device performance and lifetimes due in part to excessive heating during operation. Although substrates such as silicon or copper would be more ideal, integration by direct deposition and fabrication of III-nitride-based laser devices on these materials are either impracticable or result in poor-quality devices. Similarly, the integration of InGaN-based blue light-emitting diodes (LEDs) on low-cost materials such as Si, glass or polymers suffers from substantial sacrifices to the microstructural quality and device performance. Integration of these disparate materials systems though would allow cost-effective applications in color displays, high-resolution laser printers, high-density optical storage devices, high-power electronics, bioanalytical microsystems and room lighting based on the III-nitrides materials system. A viable alternative integration method to preclude the direct deposition of the III-nitrides onto disparate substrate materials is through wafer bonding and thin-film lift-off processes. For example, a fully functional InGaN-based blue LD originally fabricated and optimized on its growth sapphire substrate could be separated and transferred onto another host material such as copper or Si in order to further enhance the functionality of the blue LD. Such an approach allows the integration of materials selected and pre-fabricated exclusively for optimal device performance rather than for materials growth compatibility.

LASER LIFT-OFF AND WAFER BONDING

Thin-film laser lift-off (LLO) techniques have recently been established as an effective tool for the integration of GaN thin films with dissimilar materials, eliminating the sapphire substrate constraint. Separation of GaN thin films from sapphire substrates by laser processing was first demonstrated using the third harmonic of a Q-switched Nd:YAG laser with the incident beam directed through the transparent sapphire.[2] A two-step LLO process utilizing a KrF pulsed-excimer laser followed by a low-temperature (~ 40°C) separation step has also been demonstrated to successfully separate and transfer GaN thin films from sapphire onto Si, GaAs and polymer substrates without degradation to the structural and optical quality of the thin film.[3-5] Recent reports from various groups have also demonstrated the fabrication of free-standing GaN-based blue light-emitting diodes (LEDs),[6] vertical-structure blue LEDs,[7,8] blue LEDs on Si,[9] and large-area free-standing GaN substrates[10] by LLO. The robustness of the two-step LLO process was established with the fabrication of free-standing (In,Ga)N-based blue LEDs, pre-fabricated on sapphire substrates, without degradation to the device performance after laser processing.[6]

More recently, a transient liquid-phase (TLP) Pd-In wafer-bonding process was demonstrated for joining GaN/sapphire structures onto Si, GaAs, and polymer substrates.[5] In this process, a Pd-In bilayer is deposited onto the GaN surface that is then bonded onto a Pd coated Si, GaAs or polymer receptor substrate. Unlike conventional wafer-bonding methods, which typically require an ultra-clean and smooth surface, the Pd-In TLP wafer-bonding process uses a liquefied indium interface to fill in and around sub-micron surface asperities.[11] By starting with a 1:3 Pd:In ratio, the formation of a $PdIn_3$ intermetallic compound at 200°C is formed in which the molten indium layer is consumed by reacting with Pd. The resulting $PdIn_3$ compound has a melting point of 664°C. This wafer bonding approach allows for the maximum flexibility for a wide range of materials combinations for integration of InGaN-based optoelectronics.

InGaN LIGHT-EMITTING DIODES ON Si

Light-emitting diode device structure

A demonstration of the efficacy of the low-temperature TLP bonding and LLO integration methodology was performed using InGaN single-quantum-well (SQW) LEDs. The InGaN SQW structures were grown by metalorganic chemical-vapor deposition (MOCVD) on c-face sapphire substrates. First, a 4 μm thick Si-doped GaN layer was deposited followed by a 2.5nm $In_{0.25}Ga_{0.75}N$ quantum well. A 250 nm thick Mg-doped GaN layer was subsequently grown on top of the $In_xGa_{1-x}N$ SQW active region. After MOCVD growth, Ti/Au metal electrodes were deposited on the p-doped GaN:Mg layer to complete the device structure.

Blue LEDs on Si by wafer bonding and laser lift-off

A 100 nm Pd layer was then deposited onto the p-contacts by electron-beam evaporation (base pressure $\sim 1 \times 10^{-7}$ Torr) followed by a 1 μm thick In layer deposited by thermal evaporation (base pressure $\sim 5 \times 10^{-7}$ Torr). An n^{+}-Si (100) receptor substrate was separately coated with a 100 nm thick electron-beam evaporated Pd film. The Pd and In thickness were chosen such that the ratio of the Pd:In was maintained between 1:3 and 1:1 to ensure complete consumption of the In during the low-temperature bonding process. The Pd-In coated SQW LED/sapphire structures

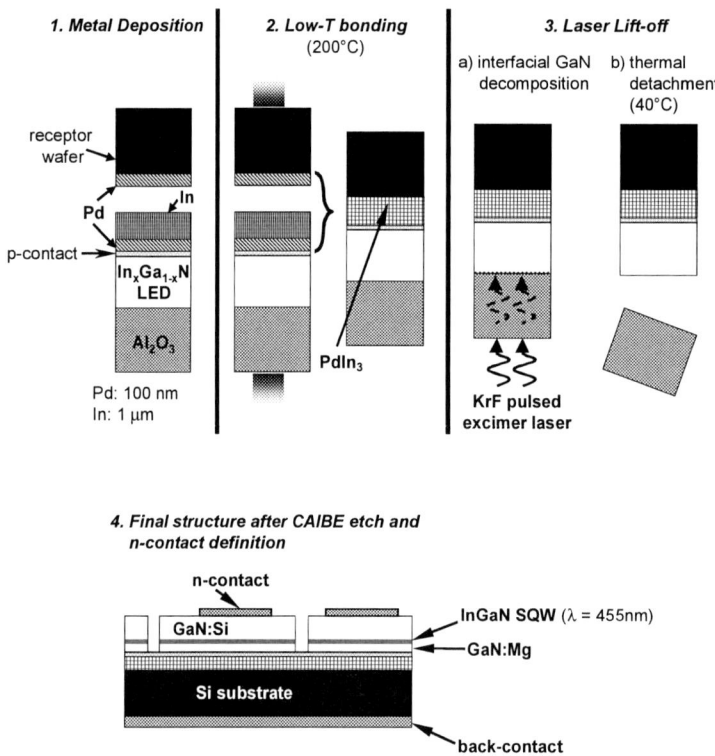

Figure 1. Process flow for bonding and transfer of InGaN SQW LED from sapphire onto Si – 1) Starting material - pre-fabricated $In_xGa_{1-x}N$ SQW LED/sapphire and Si supporting substrate; deposit Pd-In bilayer on InGaN device layer and Pd on receptor Si substrates; 2) bond $In_xGa_{1-x}N$ SQW LED/sapphire onto Si supporting substrate, 3a) KrF laser irradiation of the sapphire/$In_xGa_{1-x}N$ SQW LED/PdIn$_3$/Si structure through the transparent sapphire substrate, 3b) heat post-laser-processed structure above melting point of Ga to release sapphire substrate, and 4) CAIBE etch to isolate device and deposition of n-contact metal. The inset shows $In_xGa_{1-x}N$ SQW structure after LLO and metal contact definition.

were then bonded under a pressure of ~2.8 MPa onto the Pd-coated n^{+}-Si substrate at 200°C.[5]

The LLO processing of the sapphire/SQW-LED/Pd-In/Si structure was performed in air using a KrF pulsed excimer laser (38 ns pulse width). The decomposition of the interfacial GaN into Ga metal and N_2 gas was accomplished with a single 600 mJ/cm^2 laser pulse directed through the transparent sapphire substrate, irradiating an area of 0.03 cm^2. The laser pulse was stepped over an area of 1 cm^2 to complete the interface decomposition. By melting the thin metallic Ga interfacial layer (T$_m$ = 30°C) after laser irradiation, lift-off and transfer of the GaN film from sapphire onto the Si substrate was completed. A thin Ga-rich layer on the surface of the exposed interface was easily removed with a 1:1 solution of HCl and de-ionized water. The bonding and transfer process is shown in Figure 1.

Approximately 2 µm of the top GaN:Si layer was then removed by chemically assisted ion beam etching (CAIBE) and subsequent grooves were etched into the GaN:Mg layer to isolate an array of individual 250×250 µm^2 LED devices.[12] Finally, 100µm diameter Ti/Al metal contacts were deposited on the exposed n-doped GaN:Si layer forming an inverted vertically-connected LED structure with the Si wafer acting as the p-backside contact.[13] The combined TLP wafer-bonding and LLO processes were able to reproducibly fabricate an array of 250×250 µm^2 devices on Si over a 1×1 cm^2 area, which were then characterized by current-voltage (I-V) and electroluminescence measurements.

Device characteristics of LEDs on Si

Figure 2 shows the I-V characteristics of a typical LED device on Si. The turn-on voltage for these devices was measured to be ~ 2.5 Volts with a forward current of 100 mA at ~5.4 Volts. The turn-on voltage is somewhat higher compared to commercial LEDs on sapphire[14] and could be due partly to the non-optimized p- and non-alloyed n-contacts used for this experiment. Nevertheless, the non-optimized devices show significantly improved I-V characteristics compared to recently reported vertically-connected GaN-based LEDs on Si fabricated by direct growth.[15-17]

Room-temperature emission spectra of a typical InGaN SQW LED measured at pulsed and dc forward currents are shown in Figure 3. At a forward current of 50 mA, an emission peak

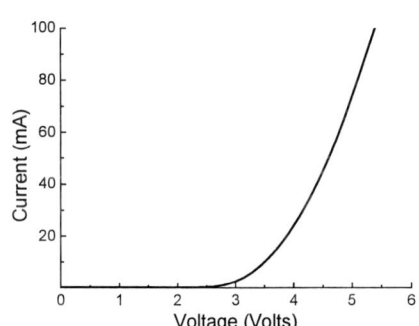

Figure 2. Room-temperature DC current-voltage characteristics for a typical 250 µm × 250 µm In$_x$Ga$_{1-x}$N SQW LED on Si.

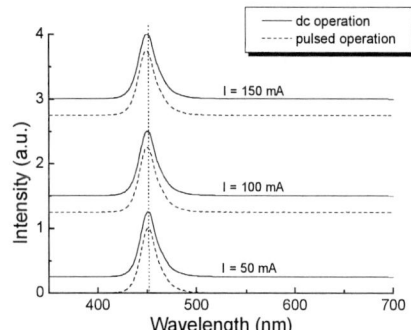

Figure 3. Room-temperature emission spectra for the In$_x$Ga$_{1-x}$N SQW LED measured at 50, 100, and 150 mA forward current for dc (solid lines) and pulsed (dashed lines) operation.

at 455 nm with a full-width at half maximum (FWHM) of ~19 nm (ΔE = 114 meV) was measured. Below 50 mA no measurable shift of the emission peak or change in the FWHM was observed. At higher currents (I = 100 mA), the emission peak exhibits a blue-shift of ~2 nm, which we attribute to band filling within the $In_xGa_{1-x}N$ SQW.[18] It should be noted, that the wavelength shift for all currents was identical whether the devices were operated under pulsed or dc conditions. This result indicates that heating of the device, when operated under dc conditions, was largely avoided since device heating would result in a red-shift of the emission spectra. In comparison to conventional LEDs on sapphire substrates, which were fabricated from the same (In,Ga)N heterostructures, a net red-shift of ~2 nm at high currents (I = 100 mA) was observed. We attribute the reduced thermal resistance for the $PdIn_3$/Si structure to the higher thermal conductivity of the Si substrate in comparison to sapphire.

InGaN MULTIPLE-QUANTUM-WELL LASER DIODES ON COPPER AND DIAMOND SUBSTRATES

Laser-diode device fabrication

The transfer of multiple-quantum-well MQW LDs from sapphire onto Cu and diamond substrates further demonstrated the robustness of the LLO process. The laser structures were grown on sapphire substrates by metalorganic chemical vapor deposition (MOCVD). First a 2 μm thick GaN film was deposited on a (0001) c-plane sapphire substrate followed by a 100 nm thick silicon dioxide (SiO_2) layer. The SiO_2 film was subsequently patterned to form an 8 μm wide stripe pattern with a period of 11 μm parallel to the GaN <10$\bar{1}$0> direction. MOCVD growth was then resumed with a 15 μm-thick, Si-doped GaN layer deposited onto the patterned substrate. The growth was completed with a standard laser diode heterostructure as follows:[19] a 100 nm thick Si-doped InGaN defect reducing layer, a 1 μm thick Si-doped AlGaN/GaN strained-layer-superlattice cladding layer, an active region comprised of three 3.5 nm thick InGaN quantum wells sandwiched between 100 nm thick GaN waveguiding layers, a 20 nm Mg-doped AlGaN tunnel barrier layer grown on top of the MQW, a 500 nm thick Mg-doped AlGaN cladding layer, and finally a 50 nm thick Mg-doped GaN contact layer.

After MOCVD growth, 2 μm ridge-waveguide structures were formed by etching into the AlGaN cladding layer with chemically-assisted ion beam etching (CAIBE). Mirrors were fabricated also with CAIBE. Metal contacts were then deposited on the top p-type GaN layer through openings in a silicon oxynitride overlayer. Finally, a highly reflective dielectric coating (R=90%) was deposited on the backside facet in order to minimize the mirror loss.

Integration of blue lasers on copper substrates

The InGaN MQW LD structures were then bonded onto the surface of a boron doped, p-type Si (100) wafer by using an ethyl cyanoacrylate (C_6-H_7-NO_2)-based adhesive, to form a sapphire/LD/adhesive/Si structure. Lift-off and transfer of the LD structures from sapphire onto the receptor Si substrate were accomplished by using a single 20 ns, ~500 mJ/cm² pulse from a XeCl pulsed-excimer laser (308 nm) directed through the transparent sapphire substrate. After the laser irradiation, a low-temperature (40°C) anneal completed the separation process by melting the Ga-rich interface.

In order to improve the structural rigidity and minimize the mechanical failure of the LD membranes and the facet mirror coatings, a secondary-supporting layer consisting of a 5 μm thick indium film was deposited onto the exposed GaN interface before removal of the support substrate. Immersion of the In/LD/adhesive/Si structure in acetone, to dissolve the adhesive bond, then completed the fabrication of 1 cm^2 free-standing membranes. The membranes were then transferred and bonded at ~200°C onto a Cu or diamond substrate by using the indium layer first as a structural support and subsequently as a bond interface. The process flow is shown in Figure 4.

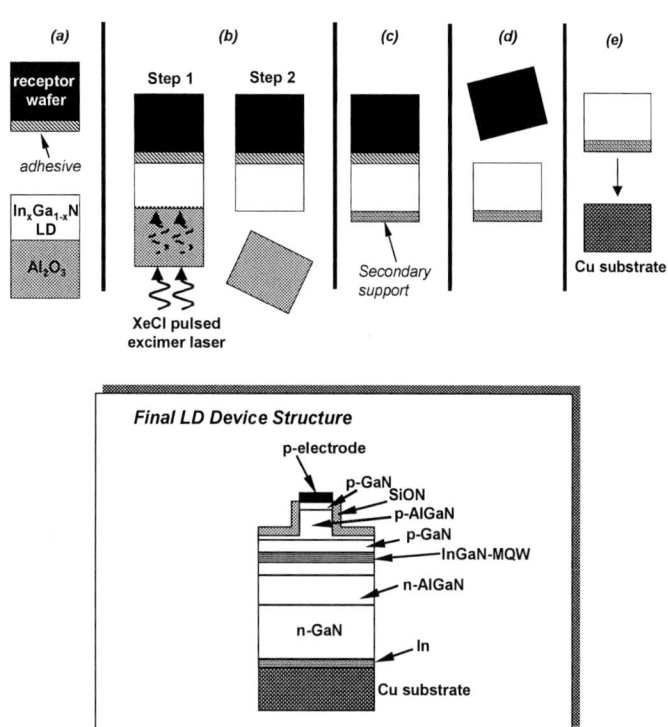

Figure 4. Process flow for fabricating an InGaN MQW laser diode on Cu - (a) Starting material: pre-fabricated InGaN MQW LD/sapphire and Si supporting substrate; (b) Step 1: Bond InGaN MQW LD/sapphire onto Si supporting substrate and irradiate the sapphire/InGaN MQW LD/adhesive/sapphire structure through the transparent sapphire substrate with an excimer laser, Step 2: heat post-laser processed structure above melting point of Ga to release sapphire substrate; (c) Deposit secondary support layer; (d) Immerse secondary support/LD/adhesive/Si structure in solvent to release LD membrane; (e) Bond membrane to Cu substrate; (inset) The final InGaN MQW LD structure on the copper substrate.

An advantage to removing the sapphire growth substrate is the ability to easily cleave mirror facets on the LDs before transfer.[20] Figure 5 shows a cross-sectional scanning electron microscopy view of a cleaved mirror facet for a typical ridge wave-guide LD. The micrograph shows a very smooth topography along the surface of the facet with a surface roughness less than 0.5 nm measured by atomic force microscopy over a 1×1 μm^2 area. The surface roughness was an order of magnitude smoother than typical CAIBE etched facets.

Figure 5: Cross-sectional SEM of a cleaved mirror facet on a free-standing InGaN ridge waveguide laser diode.

Device characteristics of transferred blue laser diodes

Characterization of these devices showed no measurable electrical or optical degradation after the integration process.[21,22] Figure 6 is a plot of the light output versus current and voltage (L-I-V) characteristics for a typical 2×800 μm^2 ridge waveguide device under room-temperature continuous-wave (cw) operation after transfer from sapphire onto diamond. The threshold current for a transferred device on diamond was 87 mA at 5.2 V, with a 0.5 W/A differential slope efficiency. These devices showed no measurable degradation after the laser processing, which suggests that the optical properties of the transferred LDs were unaffected, and damage to the waveguide due to micro-cracking or deterioration of the mirror facet coatings had not occurred.

The effectiveness of using a thermally and electrically conductive Cu substrate to improve InGaN-based LD performance is demonstrated in the LD output-power performance. Figure 7 shows the L-I curve for a typical 2×500 μm^2 LD on sapphire compared to a LD on Cu with a backside n-contact. By using the Cu substrate as an efficient heat sink, it was possible to reduce the thermal resistance of the laser diode and raise the cw light output to more than 100 mW.[20] The enhanced diode performance further establishes the use of LLO as a tool for integration allowing one to independently optimize the growth conditions and GaN-based optoelectronic device performance that can later be combined with virtually any substrate platform.

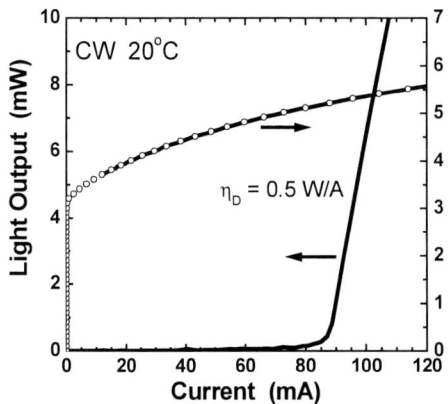

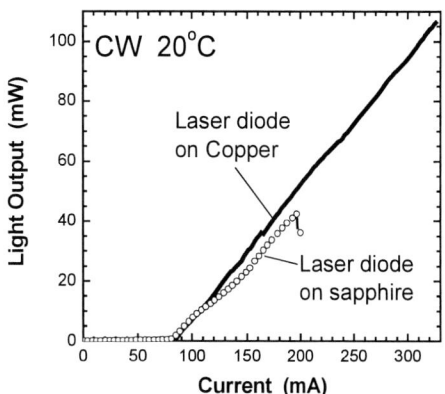

Figure 7: CW L-I-V characteristics for a 2 μm ridge waveguide LD on diamond.

Figure 6: L-I characteristics for a typical LD on Cu (solid line) compared to a LD on sapphire (dotted line). Due to the lower thermal resistance the maximum light output power is significantly improved for laser diodes on Cu substrates.

Conclusion

A LLO and transfer technique was demonstrated to integrate InGaN-based blue light-emitting and blue laser diodes with dissimilar substrates such as silicon and copper. The integration process effectively removed the sapphire growth substrate from the thin-film device allowing the transfer of the thin-film membrane onto another host substrate. The device performance of the blue LEDs on Si was comparable to those of commercially available LEDs on sapphire. Laser diodes transferred onto Cu or diamond substrates had improved device performance with a two-fold increase of the light-output after transfer onto Cu platforms. The successful transfer of the blue LEDs and LDs further demonstrate the efficacy of the LLO process to integrate InGaN-based optoelectronics onto virtually any substrate material.

References

1. S. Nakamura, M. Senoh, S. Nagahama, N. Iwasa, T. Yamada, T. Matushita, Y. Sugimoto, T. Kozaki, H. Umemoto, M. Sano, and K. Chocho *Jpn. J. Appl. Phys., Part 2* **37** L627 (1998).
2. M.K. Kelly, O. Ambacher, R. Dimitrov, R. Handschuh, and M. Stutzmann, *Phys. Stat. Sol. (A)* **159** R3 (1997).
3. W.S. Wong, T. Sands, and N.W. Cheung, *Appl. Phys. Lett.* **72** 599 (1998).
4. W.S. Wong, J. Krüger, Y. Cho, B.P. Linder, E.R. Weber, N.W. Cheung, and T. Sands, *Proceedings of the Symposium on LED for Optoelectronic Applications and the 28th State of the Art Programs on Compound Semiconductors* **98-2** 377 (1999)

5. W.S. Wong, A.B. Wengrow, Y. Cho, A. Salleo, N.J. Quitoriano, N.W. Cheung, and T. Sands, *J. Electron. Mater.* **28** 1409 (1999).
6. W.S. Wong, N.W. Cheung, T. Sands, M. Kneissl, D.P. Bour, P. Mei, L.T. Romano, and N.M. Johnson, *Appl. Phys. Lett.* **75** 1360 (1999).
7. M.K. Kelly, O. Ambacher, R. Dimitrov, R.H. Angerer, R. Handschuh, and M. Stutzmann, *Mat. Res. Soc. Symp. Proc.* **482** 973 (1998).
8. Y.K. Song, M. Diagne, H. Zhou, A.V. Nurmikko, C. Carter-Coman, R.S. Kern, F.A. Kish, and M.R. Krames, *Appl. Phys. Lett.* **74** 3720 (1999).
9. W.S. Wong, N.W. Cheung, T. Sands, M. Kneissl, D.P. Bour, P. Mei, L.T. Romano, and N.M Johnson, *Appl. Phys. Lett.* **77**, 2822 (2000).
10. M.K. Kelly, R.P. Vaudo, V.M. Phanse, L. Görgens, O. Ambacher, and M. Stutzmann, *Jpn. J. Appl. Phys.* Part 2 **38**, L217 (1999).
11. B.J. Dalgleish, K. Nakashima, M.R. Locatelli, A.P. Tomsia, and A.M. Glaeser *Ceramics International* **23**, 313 (1997).
12. M. Kneissl, D. Hofstetter, D.P. Bour, R. Donaldson, J. Walker, and N.M. Johnson, *J. Cryst. Growth* **189/190**, 846-849 (1998).
13. M. Kneissl, D.P. Bour, B.S. Krusor, L.T. Romano, N.M. Johnson, M. McCluskey, W. Goetz, R.D. Bringans, *SPIE Proceedings* **3279**, 69-76 (1998).
14. S. Nakumura, M. Senoh, N. Iwasa, and S.I. Nagahama, *Appl. Phys. Lett.* **67**, 1868 (1995).
15. S. Guha and N.A. Bojarczuk, *Appl. Phys. Lett.* **72**, 415 (1997).
16. C.A. Tran, A. Osinski, R.F. Karlicek, Jr., and I. Berishev, Appl. Phys. Lett. **75**, 1494 (1999).
17. J.W. Yang, A. Lunev, G. Simin, A. Chitnis, M. Shatalov, M. Asif Khan, J.E. Van Nostrand, and R. Gaska, *Appl. Phys. Lett.* **76**, 273 (2000)
18. S. Nakamura, J. Vac. Sci. Technol. A **13**, 705 (1995).
19. M. Kneissl, D.P. Bour, L.T. Romano, C.G. van de Walle, J.E. Northrup, W.S. Wong, D.W. Treat, M. Teepe, T. Schmidt, N.M. Johnson, *Appl. Phys. Lett.* **77**, 1931 (2000).
20. M. Kneissl, W.S. Wong, D.W. Treat, M. Teepe, N. Miyashita, N.M. Johnson, submitted to *IEEE J. Sel. Top. Quantum Eletron.* (2001).
21. W.S. Wong, M. Kneissl, P. Mei, D.W. Treat, M. Teepe, and N.M. Johnson, *Appl. Phys. Lett.* **78**, 1198 (2001).
22. W.S. Wong, M. Kneissl, P. Mei, D.W. Treat, M. Teepe, and N.M. Johnson, *Jpn. J. Appl. Phys., Part 2* **39**, L1203 (2000).

Mat. Res. Soc. Symp. Proc. Vol. 681E © 2001 Materials Research Society

Ultrathin Slices of Ferroelectric Domain-Patterned Lithium Niobate by Crystal Ion Slicing

David A. Scrymgeour[1] and Venkat Gopalan[1], Tony E. Haynes[2], and Miguel Levy[3]
[1]Materials Research Laboratory, Pennsylvania State University,
University Park, PA 16802;
[2]Solid State Division, Oak Ridge National Laboratory, Bldg. 3003, MS-6048, P. O. Box 2008, Oak Ridge, TN 37831
[3] Physics Department
Michigan Technological University, Houghton, MI 49931

ABSTRACT

We report the successful fabrication of 6 μm thick slices from a ferroelectric domain micro-engineered $LiNbO_3$ wafer device using the crystal ion slicing technique. The device was created by micropatterning ferroelectric domains in a bulk 0.3 mm thick wafer of z-cut $LiNbO_3$, followed by ion-implanting with 3.8 MeV He^+ ions to a fluence 5 x 10^{+16} ions/cm^2 to create a damage layer at a well defined depth from the surface. Etching away this damaged layer in dilute hydrofluoric acid results in a liftoff of the top slice in which the ferroelectric domain patterns are left intact. The influence of annealing conditions on liftoff time and depth of etch lines was studied. Helium-Neon laser light was successfully coupled into the device. Due to unintentional breakage of the polished input and output faces, the electro-optic scanning performance has not been characterized so far.

INTRODUCTION

The ability to control the angular position of a laser beam with high speed is of interest in many applications including optical communications, optical data storage, laser printing, and display technologies. Active solid-state electro-optic scanners based on micro-patterned $LiNbO_3$ have several advantages over mechanical and other systems including small device size and high operating speed (intrinsic response frequencies >100GHz). However, widespread application of these devices is currently limited because of the large voltage required to operate devices fabricated in single crystal wafers. A solution to this problem is to make the devices thinner, thereby reducing the operating voltage for a fixed driving electric field. However, the fabrication of high-quality thin films of $LiNbO_3$ has so far been elusive. Techniques such as pulsed laser deposition[1] and chemical vapor deposition[2] yield polycrystalline films, which have electro-optic responses much smaller than single crystal bulk $LiNbO_3$. Further, the light propagation losses in thin films are still high, which makes them unsuitable for optical devices.[3]

Recently, an ion-implantation-based technique called crystal ion slicing (CIS), has been reported for making thin slices of ferroelectric lithium niobate.[4,5] CIS allows a free standing micron-thick single crystal film to be fabricated from a bulk crystal. In CIS, ion implantation is used to define a damaged sacrificial layer several microns below the surface. This sacrificial layer is then etched away by immersion in a dilute hydrofluoric acid solution. The large etch selectivity of the damage layer over the rest of the crystal allows for the lifting off of a thin film, whose thickness is well controlled by the implantation energy.

By crystal ion slicing of domain-engineered devices fabricated on wafers, operating voltages can be significantly reduced. In this paper, we show that indeed domain-micropatterned devices can be sliced into thin layers without destroying the domain patterns defined in the original crystal. We first describe the principle and fabrication of an electro-optic device used for laser beam scanning, followed by a description of the CIS processing of the fabricated device.

Electro-optic Scanner device on LiNbO₃

The most basic geometry for deflecting light is by use of an optical prism. In paraxial approximation, where the deflection angles are small, one can use a series of prisms in sequence, each prism successively deflecting the light beam further and further from the optical axis as shown in Fig. 1.

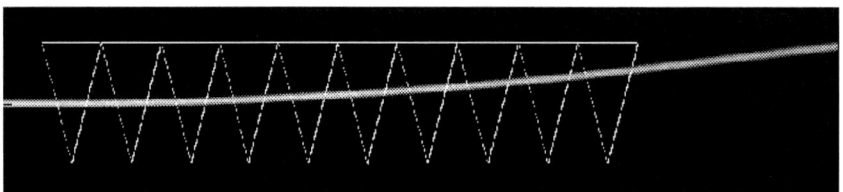

Figure 1: A beam propagation method (BPM) simulation of a rectangular scanner in LiNbO₃ operating at +15 kV/mm. A total of 10 triangles is shown, each triangle is 150 μm wide by 900 μm high. The total deflection is 4.3° from center, which gives 8.6° total deflection for ± 15 kV/mm.

The simplest scanner design is a rectangular scanner geometry which consists of N identical prisms placed in sequence, each with a base l and height W, such that the total length of the rectangular scanner is L=Nl, and width is W. The total deflection angle, θ_{int}, at the output for a light beam incident along the axis, L, of the rectangular scanner is given by[6]

$$\theta_{int} = \frac{\Delta n}{n} \frac{L}{W} \tag{1}$$

where n is the index of the material, and Δn is the index difference between the prism and the surrounding matrix, where $\Delta n \ll n$. These prisms are fabricated in an electro-optic medium where the refractive index n, and therefore the Δn is electric field tunable. This results in an electric field controlled deflection angle θ_{int}.

In a uniaxial electro-optic crystal such as LiTaO₃ or LiNbO₃, the refractive indices are electric field tunable. If the uniaxial direction is denoted subscript 3, then a decrease in the refractive extraordinary index n_e (or n_3) with the application of an electric field E_3 parallel to the spontaneous ferroelectric polarization direction is given by $\Delta n_e = -\frac{1}{2} n_e^3 r_{33} E_3$. If ferroelectric domains were created in the shape of a series of prisms in a z-cut crystal of LiNbO₃ or LiTaO₃, one can create an electro-optic scanner. Without an electric field, the refractive indices inside and outside the domain prisms will be equal, thus resulting in no deflection of light. When a uniform electric field is applied across the z-cut crystal in the z-direction, the electric field E_3 is parallel to the spontaneous polarization in one domain and antiparallel in the other. This results in an index difference in the extraordinary index of $(n_{prism}-n_{matrix}) = +/- 2\Delta n_e$ across the ferroelectric domain walls defining the prisms. Such rectangular scanners in LiNbO₃ and LiTaO₃ have been demonstrated.[7,8,9]

In LiNbO₃ the coercive field strength (the field required to reverse the domain at room temperature), E_c, is ~21 kV/mm.[10] So in a scanner as previously proposed, the maximum device operating field strength should be less than the coercive field strength, so as to not destroy the domain patterns. On the other hand, the field should be as high as possible to achieve maximum deflection. For a typical field strength of 15 kV/m, if the device was fabricated in a commercially available crystal of thickness 0.3 mm, this would mean that the device would operate at (15 kV/mm)(0.3 mm) = 4500 V. This high operating voltage is a major hurdle to broader application of this technology, since electrical drivers that can provide large voltages at high bandwidths of GHz are not practical. A solution to this is to make the crystal thinner, so as to reduce the operating voltage.

EXPERIMENTAL DETAILS

A bulk device was fabricated on a 300 ± 5 μm thick, Z-cut single crystal, single-domain $LiNbO_3$ wafer. Domain micropatterning was performed as follows. A photoresist pattern resembling the device shown in Fig. 1 was defined on the +z face of the crystal surface with conventional contact photolithography. A uniform Tantalum (Ta) electrode was then deposited on the sample by DC sputtering. This was followed by liftoff process, which removes the metal in areas where photoresist exists beneath, by dissolving the underlying photoresist pattern in acetone. This leaves behind a complementary Ta-film pattern on the crystal surface resembling Fig. 1, with the Ta defining the complement of the area of the prisms composing the scanner. A photoresist coating is then deposited uniformly on the surface of the sample, across both the electrode and crystal features and baked well to remove any solvents. This layer has been found to improve selectivity of the domain reversal process by suppressing domain reversal beyond electrode edges through reduction of surface conduction in areas without the electrode. For domain inversion, ~21kV/mm electric field is applied across the crystal. The positive electrode is the Ta-film pattern on +z face while a uniform water electrode is used on the -z face as the ground. After domain reversal, the metal electrode is then removed by etching in HF.

This bulk device with patterned domain is then ion implanted with a dose of $5 \times 10^{+16}$ ions/cm^2 at an energy of 3.8 MeV with the temperature of the sample kept below 12°C during implantation. The input faces of the implanted devices are then end polished to 0.05 μm finish. After polishing, optimal condition rapid thermal annealing at 450°C for 40 s in flowing forming gas (5% H$_2$, 95% Ar) was used to enhance the etch selectivity between sacrificial layer and the rest of the sample. The device was then encapsulated in an HF resistant wax, which was used to protect the film's top surface and polished ends. Finally the device is immersed in dilute HF (10%) until it is completely detached. After detachment, electrodes were applied on each face with sputtered gold, and the sample (or sample fragments) were mounted and electrical contacts made to the surfaces. A beam of He-Ne laser light was focused to 2.5 μm in the vertical and 100 μm in the horizontal and targeted on the input faces of the device. The output face was imaged by a microscope objective and the light collected in a CCD camera.

DISCUSSION

The intermediate annealing step in the liftoff process above, allows relatively large sample areas (2 mm x 10 mm) to be lifted off in ~32 hours. This is believed to occur by promoting the diffusion of helium into the implantation induced micro voids and the buildup of an internal pressure in the sacrificial layer. Changing the annealing temperature and atmosphere had an effect upon both the time required for lift off and the quality of the lifted off slice. In small test pieces (1.5 mm x 5 mm) of LiNbO$_3$ implanted at the same energy some general trends were noticed. In general, annealing temperatures lower than 450°C took longer to lift off as shown in Figure 2(a). For example, the lift-off times were 6 hours at 450°C, ~12 hours for 400°C, and >24 hours with no annealing. Annealing temperatures higher than 450°C tended to cause spalling. Other annealing atmospheres, such as Ar, N$_2$ and air, caused longer lift off times. The liftoff face showed etch lines that were characterized by atomic force microscopy as shown in Figure 3. The optimal lift conditions yielded an etch depth of 20.2 nm, which was much better than >40 nm for other processing conditions. The density of the etch lines was found to have no clear relation between process atmosphere, temperature, or exposure time to acid.

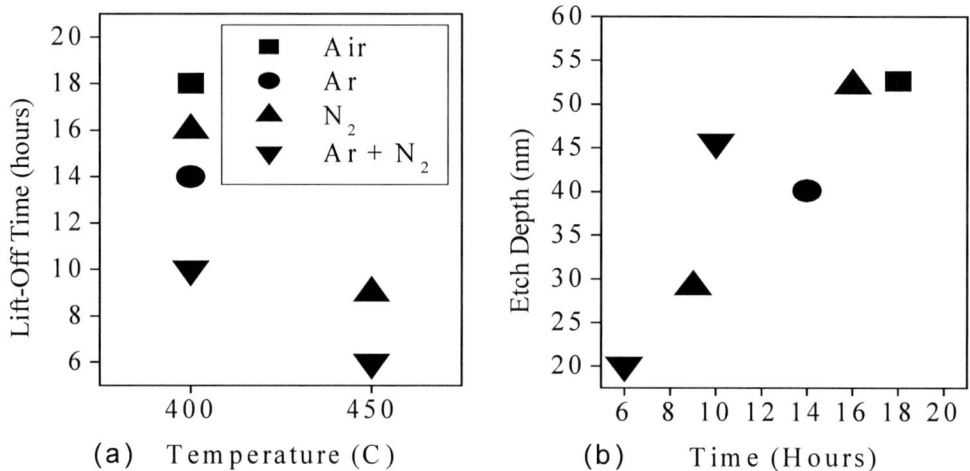

Figure 2: (a) Lift off time increased for higher processing temperatures. (b) Etch depth was found to scale linearly with exposure to acid

A CIS domain engineered device 2 mm x 9 mm was successfully lifted off using optimal conditions. Figure 4 shows a composite optical micrograph of the device. The domain structures patterned in the crystal before the implantation, annealing, and lift off step are still present in the crystal. The crystal is 6 μm thick, which was determined by imaging the crystal edge under a microscope. Figure 4 also shows some of the inherent difficulties in lifting off large size devices. Notice the large chip out of the input face and damage to the output face. If these chips are in the entrance or exit path for the scanner channel, no light could be coupled. Also, the lifted off scanners tended to show curvature in the plane. This curvature caused cracking and breaking of the sample in subsequent processing steps, like electrode coating and mounting. Because of these problems, only partial device fragments were mounted and tested in this study. Light from a He-Ne laser was successfully coupled into a fragment with a polished output face and an input face which was unintentionally fractured. Figure 5 shows the light exiting the output face, as imaged by a Charge Coupled Device(CCD) camera. However, since the light was not in a scanner channel in this region, the device scanning could not be tested in this work. However, electro-optic modulation has been previously reported in similar CIS sliced pieces, and we believe these devices would work with better input and output faces.[11]

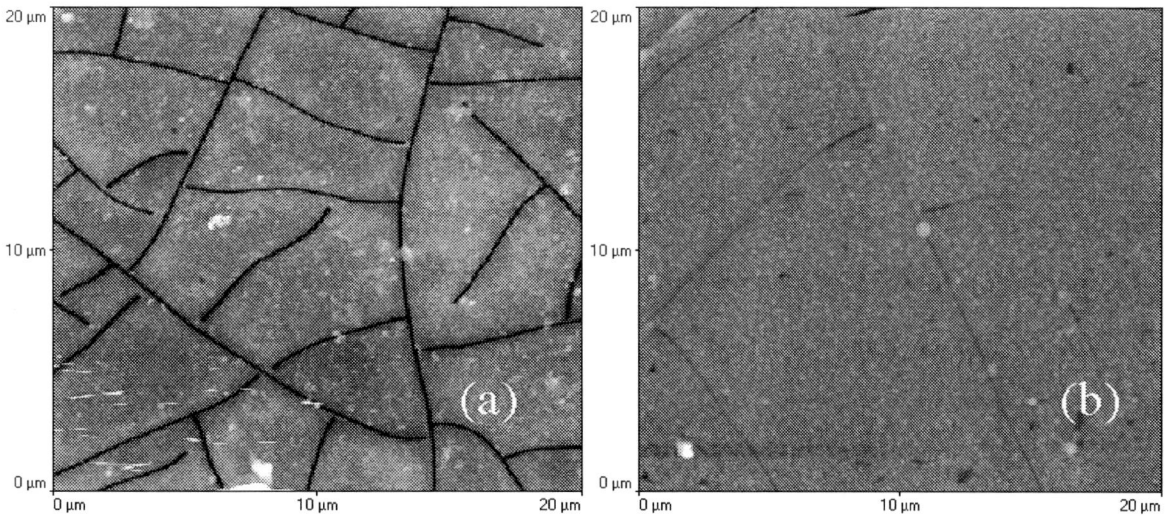

Figure 3. AFM images (a) 400°C in N_2 lift off for 16 hours with etch depth of 50.52 nm. (b) 450°C in forming gas lift off for 6 hours with etch depth of 20.31 nm.

Figure 4: A top view of a CIS electro-optic scanner on z-cut $LiNbO_3$ The triangles are oppositely oriented domain states with a base of 1000 µm and height of 775 µm. The first triangle has been enhanced to show location. The sample curves up the right hand side.

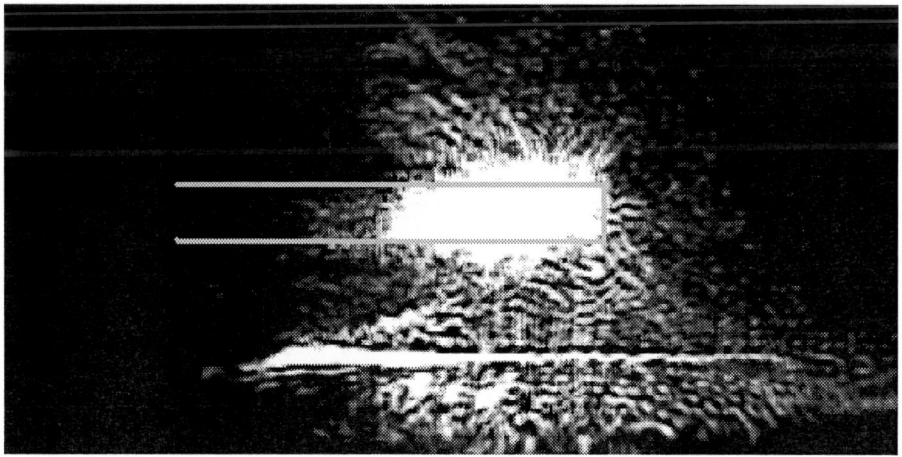

Figure 5: The output face of the crystal imaged in a CCD camera showing light coupling. The gray line shows position of device.

CONCLUSIONS

The successful lift off of a 2mm x 9 mm microengineered electro-optic device in $LiNbO_3$ was achieved by crystal ion slicing. The device slice measures ~6 µm thick. The predefined prismatic domain structures were preserved in the final device. Theoretically this device should operate at ± 90 V, which is significantly lower than the ± 4500 V required for a 0.3 mm thick bulk device. Optimal conditions for pre-liftoff annealing at 450°C for 40 s in flowing forming gas (5% H_2, 95% Ar) was found to have the shortest liftoff time and produced minimal etching of surface (~20 nm depth of etch lines). Light coupling was successfully achieved in device fragments. The scanner performance is yet to be characterized

ACKNOWLEDGMENTS

This work was partially supported by the funds made available to Dr. V. Gopalan through the National Science Foundation and the DARPA-STAB program. Work performed at the Surface Modification and Characterization Research Center of Oak Ridge National Laboratory was sponsored by the U.S. Department of Energy, Office of Science under contract DE-AC05-00OR22725 with UT-Battelle, LLC.

REFERENCES

1. Y. Shibata, K. Kaa, K. Akishi, M. Kanai, T. Kawai, and S. Kawai, *Appl. Phys. Lett.* **61**, 1000 (1992).
2. B. J. Curtis and H.R. Brunner, Mater. Res. Bull. **10**, 515 (1975).
3. Lee, S.Y.; Feigelson, R.S. *J. Crystal Gr.* Vol. 186, Iss. 4, pp. 594-606; (1998).
4. M. Levy, R.M. Osgood, R.Liu, L.E. Cross, G.S Cargill III, A. Kumar, and H. Bakhru, *Appl. Phys. Let.* **73**, 2293 (1998).
5. A. M. Radojevic, M. Levy, R.M. Osgood, Jr., A. Dumar, H. Bakhru, C. tian, and C. Evans, *Appl. Phys. Lett.* 74, 3197 (1999).
6. J. F. Lotspeich, IEEE Spectrum, **5**, 45-52 (1969).
7. K. Gahagan, V. Gopalan, J.M. Robinson, and Q. X. Jia, *Appl. Optics*, **38**, 1186-1190, (1999).
8. Q.B. Chen, Y. Chiu, D.N. Lambeth, T.E. Schlesinger, and D.D. Stancil, *J. Lightwave Technol.* **12**, 1201-1202 (1994).
9. J. Li, H.C. Cheng, M.J. Kawas, D.N. Lambeth, T.E. Schlesinger, and D.D. Stancil, *IEEE Photonics Technol. Lett.* **8**, 1286-1488 (1996).
10. V. Gopalan and M.C.Gupta, *Appl. Phys. Lett.* **68**, 1323 (1996).
11. R. A. Ramadan, M. Levy, and R.M. Osgood, Jr., "Electro-optic modulation in crystal-ion-sliced z-cut LiNbO3 thin films', *Appl. Phys. Lett.* 76, 1407 (2000).

Mat. Res. Soc. Symp. Proc. Vol. 681E © 2001 Materials Research Society

Fabrication of Three-Dimensional Photonic Crystal by Wafer Fusion Approach

Noritsugu Yamamoto[2,3], Katsuhiro Tomoda[1], and Susumu Noda[1,3]

[1]Dept. of Electronic Science and Engineering, Kyoto University,
Yoshidahonmachi, Sakyo-ku, Kyoto 606-8501, Japan
[2]Research Institute of Photonics, National Institute of Advanced Industrial Science and Technology,
AIST Tsukuba Central 2, 1-1-1, Umezono, Tsukuba 303-8568, Japan
[3]Core Research for Evolution Science and Technology (CREST), Japan Science and Technology Corporation (JST)

ABSTRACT

Based on a set of requirements identified for photonic crystals intended for use in optoelectronic devices, we have developed a method of fabricating three-dimensional photonic crystals that involves stacking air/semiconductor gratings by wafer fusion approach. Precise alignment of the stacked layers is achieved through the use of a laser beam assisted very precise alignment system, and three-dimensional photonic crystal has been successfully fabricated for the infrared and optical communication wavelength regions. We have also developed a photonic crystal waveguide providing sharp 90° bend.

INTRODUCTION

Photonic crystals, new optical materials with a periodic dielectric structure, have attracted great interest. The periodic structure of these crystals creates a periodic variation in refractive index, forming a band structure in terms of photon energy and wavevector [1-4]. The dispersion relation in the crystal has strong nonlinearity, unlike ordinary optical materials; photonic crystals are characterized by a photonic bandgap in which all electromagnetic wave propagation for all wave vectors is blocked. Various scientific and engineering applications, such as control of spontaneous emission, zero-threshold lasing, and sharp bending of light are expected to be developed using photonic bandgaps and artificially introduced defects. To make the best use of the potential of photonic crystals, the crystals should satisfy the following conditions: (i) crystals should be three-dimensional photonic crystals with a complete photonic bandgap, (ii) there should be the possibility of introducing defects at an arbitrary position or (iii) introducing light-emitting elements, and (iv) crystals should have high electrical conductivity, which is important for actual device applications. Specific functional devices, such as two-dimensional surface-emitting photonic crystal lasers and add-drop filter waveguides can be realized using two-dimensional photonic crystal membranes [4, 5], which satisfy not all of the above conditions. However, for the most advanced applications such as spontaneous emission control and ultrasmall optical integrated circuits, the crystal must satisfy the above conditions.

A number of approaches to fabricate three-dimensional photonic crystals have been proposed, including the use of self-assembled colloidal crystal (artificial opal), GaAs-based three-axis dry-etched crystal, and silicon-based layer-by-layer crystal with a woodpile structure [6-8]. However, all these techniques have disadvantages, preventing the resultant crystals from satisfying the above requirements. In this proceedings, we describe a new fabrication technique based on wafer fusion that affords crystals that satisfy all these requirements, optical properties of the crystals and fabrication of the photonic crystal with bending waveguide.

FABRICATION METHOD AND OPTICAL PROPERTIES OF THREE-DIMENSIONAL PHOTONIC CRYSTAL

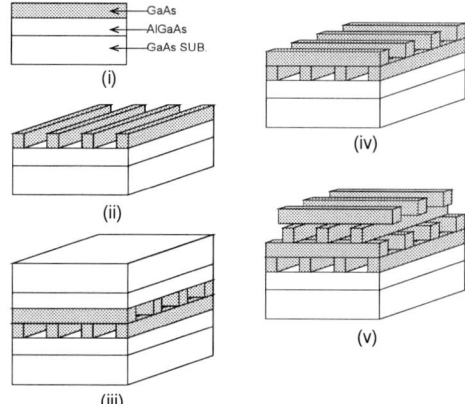

Figure 1. Schematics of fabrication method: (i) epitaxial growth, (ii) patterning of basic two-dimensional structure, (iii) wafer fusion, (iv) removal of excess substrate by selective etching, and (v) repetition of steps (iii) and (iv) using laser beam diffraction to align layers.

A schematic of the fabrication method for the three-dimensional photonic crystal is shown in Fig. 1. The method is as follows: (i) the photonic crystal layer (GaAs layer in Fig. 1(i)) and the etching stop layer (AlGaAs layer) are formed on the substrate by epitaxial growth, (ii) a basic two-dimensional structure (striped pattern in the figure) is formed using semiconductor micro-fabrication techniques such as electron beam lithography and reactive ion etching, (iii) the pair of wafers produced in step (ii) are stacked face-to-face and fused by a wafer-fusion technique, (iv) the substrate and the etching stop layer of one side of the fused wafer are etched selectively and successively, and (v) the wafer obtained in step (iv) is cleaved in two, and the wafer-fusion and selective etching steps (iii and iv) are repeated, forming a three-dimensional periodic structure[9]. By utilizing a stripe structure such as the basic two-dimensional structure shown in Fig. 1, the fabricated crystal has diamond-like symmetry and a full bandgap. Defects can be included in the crystal by changing the basic structure, e.g. enlarging part of the stripe. The other III-V semiconductors such as InP and InGaAsP can be used instead of GaAs as shown in the figure. Since III-V semiconductors are used in a wide range of photoelectronic devices, it is expected to be possible to introduce active regions using the proposed fabrication method. Furthermore, the interface formed by wafer fusion is electronically conductive and allows electric current flow, making photonic crystals fabricated by this method highly suitable for electrically controlled devices. Crystals produced by this fabrication method satisfy the four conditions described above.

Here we will describe each process in detail. Precise alignment of the stacked layers is necessary to achieve the desired diamond-like crystalline structure. As shown in Fig.1, the layers

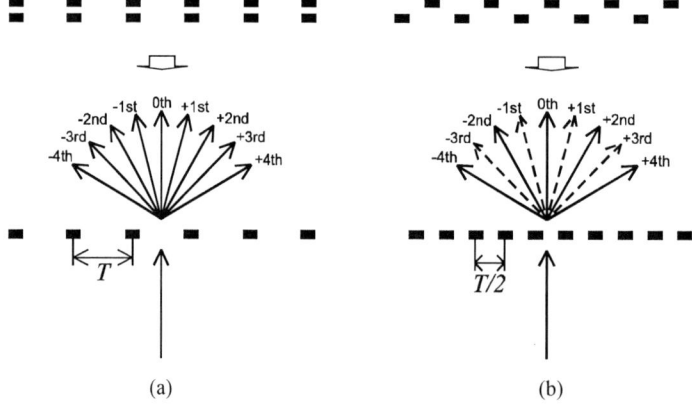

Figure 2. Principle of laser beam diffraction alignment. (a) Two gratings are in same position. (b) Two gratings offset by half a period, acting as a grating with half the period of the originals and with weaker odd-order diffractions.

Mat. Res. Soc. Symp. Proc. Vol. 681E © 2001 Materials Research Society

are stacked orthogonally, with successive parallel layers offset by half a stripe period. A precision of the order of 1/10 of a stripe period is required in order to realize the photonic crystal effect. We have developed a laser beam diffraction observation method as a means of execute such precision alignment. The principle of the technique is shown in Fig. 2. When two gratings of the same period are aligned exactly (Fig. 2(a)), they are observed as a single grating of that period. When the two gratings are offset by exactly half the period (Fig. 2(b)), they are observed as a grating with a period of half that of the original gratings and a diffraction angle of twice that for the original gratings due to destructive interference; odd-order diffractions are weakened due to the gap between the gratings. The gratings can then be aligned using the known diffraction pattern for this half-period offset. Details of the technique are stated in ref. 10.

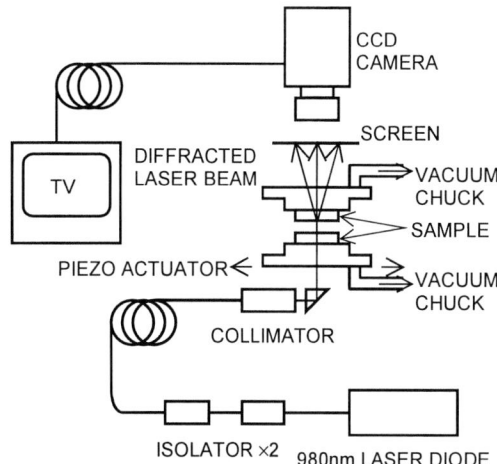

Figure 3. Schematics of alignment system based on the laser beam diffraction observation technique

Figure 3 is a schematic of the alignment system developed to align the layers of the photonic crystal. The light source used was a laser diode with a wavelength of 980 nm, at which GaAs is transparent. The wafers are fixed to the piezoactuator using a vacuum chuck, and the diffraction pattern projected onto the screen is observed by CCD camera. One of the fixtures can be moved using an XYZθ stage. To verify the accuracy of the alignment system, we aligned gratings with a period of 0.7 μm. The diffraction pattern after alignment and a cross-sectional SEM image are shown in Fig. 4. It can be seen that the gratings were aligned with a precision of less than 50 nm, which is considered to be sufficient for the fabrication of photonic crystals.

Wafer fusion is achieved as follows. The oxidized surface of the wafer is removed using a buffered HF solution. The surface is then treated with organic ammonium solution to produce a stable hydrophilic surface. Two such wafers are aligned using the alignment system described above and then pressed together by the piezoactuator at room temperature. The wafers are weakly bonded together by the hydrogen bonds of the hydrophilic surfaces. However, these room-temperature bonds are not sufficient for the bond to survive the subsequent fabrication processes, and the interface is not yet electroconductive. The wafers are therefore heated in a hydrogen atmosphere to improve the strength of the bond.

The selection of heating temperature is very important. For infrared-range crystals with a stripe period of 4 μm, the heating temperature is 700 °C. In this case, a strongly bonded interface was obtained without deformation of the stacked stripes, and a high quality bandgap effect was

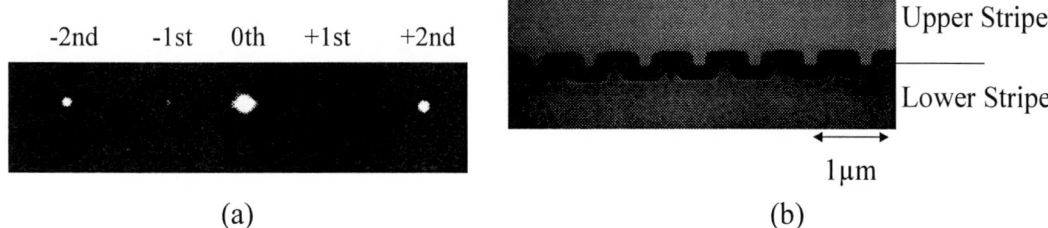

(a)   (b)

Figure 4. (a) Diffraction pattern and (b) cross-sectional SEM image after alignment. The ±1st order diffraction spots become dark. The upper and lower stripes are offset by half a period.

Mat. Res. Soc. Symp. Proc. Vol. 681E © 2001 Materials Research Society

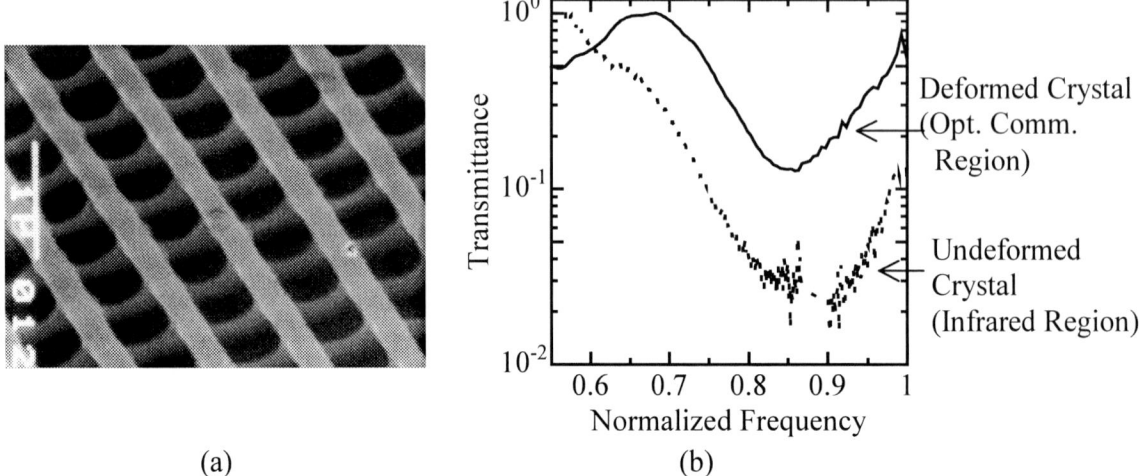

(a)   (b)

Figure 5. (a) SEM image of bonded wafer deformed by mass transport. (b) Transmission spectra of four-layer photonic crystals. Solid and dotted lines are spectra of deformed crystal for optical communication wavelength region and undeformed crystal for infrared wavelength region, respectively. Horizontal axis is normalized frequency.

obtained, as demonstrated by transmission spectrum measurements (dotted line in Fig. 5(b)). However, crystals designed for operation in the optical communication wavelength region, with a stripe period of 0.7 μm, were drastically deformed by treatment at this temperature, as shown in Fig. 5(a). The stripe thickness became thinner, particularly at the center, and the edges became rounded. The transmission spectrum of this crystal (solid line in Fig. 5(b)), shows that transmittance in the bandgap region is higher than for the non-deformed crystal [11, 12]. Therefore, the bandgap effect of the photonic crystal has been degraded. This effect is attributable to mass transport by molecular diffusion, which alters the crystal structure from the homogeneous crystalline structure created by epitaxial growth. We therefore sought the most appropriate wafer-fusion temperature, within the range 400 to 700 °C; 500~550°C was found to produce strong bonds with only limited molecular diffusion [3].

The crystal can also be degraded during the dry etching processes that are applied to remove the excess GaAs. Figure 6 outlines the removal process: (i) Selective etching of GaAs substrate by mechanical lapping and $NH_4OH+H_2O_2$ solution, (ii) Selective etching of AlGaAs etching stop layer by HF solution, (iii) Dry etching of thin GaAs layer, which protects the AlGaAs layer of opposite side wafer from HF solution at process (ii). If the dry etching time is longer than the

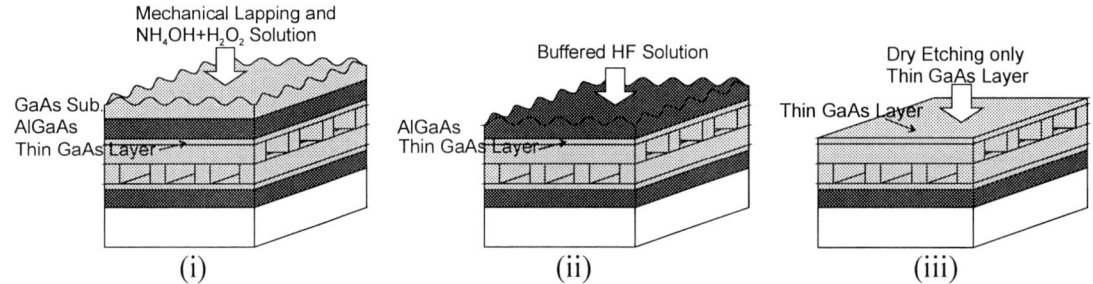

Figure 6. Process of excess substrate removal: (i) selective etching of GaAs substrate using mechanical lapping and $NH_4OH+H_2O_2$ solution, (ii) selective etching of AlGaAs layer using HF solution, and (ii) dry etching only thin GaAs layer.

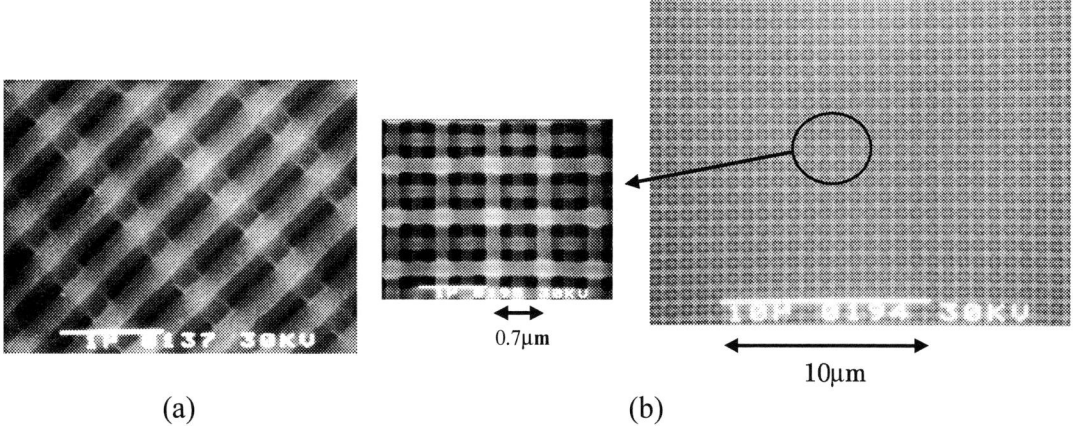

(a)   (b)

Figure 7. (a) Slant angle SEM image of two-layered crystal without deformation. (b) Top view SEM image of four-layered crystal. Well aligned, uniform and nodeformated crystal has been successfully fabricated.

optimum time, the stripes themselves will start to etch. In our dry etching system, the sample surface can be observed during the etching process. The time at etching should be ceased for an infrared photonic crystal (wide stripes) can be determined by observing the appearance of the stripe pattern directly. However, the stripe pattern of optical-communication photonic crystal is too small to make such an observation. Therefore, we judge the end of etching by observing the interference color.

Using an appropriate temperature for wafer fusion and an appropriate dry etching time according to the interference color, a deformation-free crystal was successfully fabricated, as shown in Fig. 7(a). Top view of four layered crystal fabricated by this condition is shown in Fig. 7(b). It can be seen that the fabricated crystal is very uniform. The transmission and reflection spectra of the crystal are shown in Fig. 8. The attenuation of transmission spectrum is -23dB in the bandgap, which is greater than that for crystals fabricated by previous techniques. The transmission spectra were then taken for various incident angles in order to verify that this crystal had a full bandgap. Attenuation occurs in all spectra in the 1.3 to 1.55 μm wavelength region. Reflectance was also measured, and was found to be almost 100% in this wavelength region, confirming that transmission attenuation is attributable to the bandgap effect and not absorption or diffraction loss. Further more, we have fabricated eight layer stacked photonic crystal. The transmission attenuation of the crystal is more than -40dB [3]. We think that the attenuation is enough for device application. For the first step of optical device, we fabricated the photonic crystal with bending waveguide.

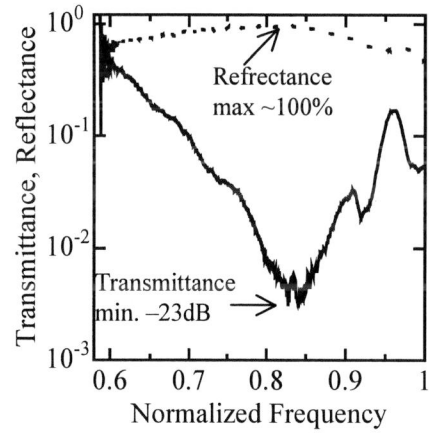

Figure 8. Transmission (solid line) and reflectance (dotted line) spectra of nondeformated four-layered photonic crystal. The transmission attenuation is enhanced rather than that of the deformed crystal shown in Fig. 5(b).

PHOTONIC CRYSTAL WAVEGUIDE

One of the most important applications of photonic crystal is in ultra-small optical integrated circuits.

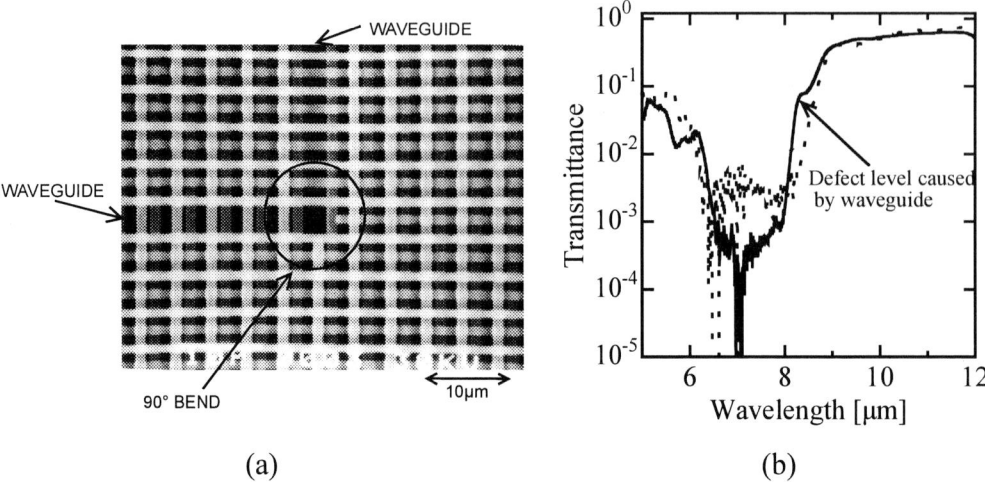

(a)   (b)

Figure 9. (a) SEM image of waveguide with 90° bend in three-dimensional photonic crystal. (b) Transmission spectrum of waveguide crystal (solid line) and eight-layered unmodified photonic crystal (dotted line)

Elements such as light emission units, electrodes for current injection, and optical waveguides are required in order to realize these circuits. Here, we have attempted to fabricate an optical waveguide that forces a sharp bend. The device was fabricated for the infrared region due to the relative ease of constructing such crystals. The linear waveguide is formed by partially removing a single stripe from two layers, causing incident light to be redirected by 90°. This waveguide structure has been shown to cause only low propagation loss between broad wavelength regions by finite difference time domain (FDTD) calculations [13].

The fabrication of this structure requires stripes to be partially removed from individual layers within the crystal. This is achieved by removing parts of stripes from the two-dimensional pattern in the formation process (Fig. 1(b)), creating waveguide layers. We fabricated a crystal having two waveguide layers between five unmodified layers (12 layers total). Figure 9(a) shows the fabricated photonic crystal with 90° waveguide before the upper five layers have been stacked [3]. The transmission spectrum for the waveguide device is shown in Fig. 9(b) together with the unmodified photonic crystal. There is a slight increase in transmission around the upper band edge for the waveguide device, considered to be the level of defect introduced by the waveguide modification. These results show that it is possible to fabricate waveguides based on three-dimensional photonic crystal. We are currently fabricating a three-dimensional photonic crystal waveguide for the optical communication wavelength region, and investigating photonic crystals with quantum wells for use as a light-emitting layer. It is expected that the integration of these developments will allow us to create optical integrated circuits.

CONCLUSION

The four requirements for optoelectronic devices based on photonic crystal were outlined, and a fabrication method for three-dimensional photonic crystals that satisfies these requirements was demonstrated. The method involves stacking air/semiconductor gratings by wafer fusion. Precise alignment of the stacked gratings was achieved by using an alignment system developed for this purpose, in which gratings are aligned based on the observation of laser beam diffraction patterns. Precision of better than 50 nm was achieved using this technique. Deformation problems associated with mass transport during heat treatment in the wafer fusion process were resolved by choosing an appropriate treatment temperature; 500 °C for air/GaAs gratings. Degradation due to over-etching in the dry-etching process following wafer fusion was prevented by ceasing etching according to an observed change in interference color. Using all these techniques, we successfully fabricated a three-dimensional photonic crystal with strong photonic bandgap effect in a wavelength region suitable for optical communication. As the first step toward developing an optical integrated circuit, we fabricated a three-dimensional photonic crystal waveguide with a 90° bend for the infrared wavelength region. We are currently investigating a waveguide for the optical communication region and the incorporation of a light-emitting layer in these crystals. We believe that this technology will lead to the realization of ultra-small optical integrated circuits.

REFERENCES

1. E. Yablonovitch, *Phys. Rev. Lett.* **58**, 2059 (1987)
2. S. John, *Phys. Rev. Lett.* **58**, 2486 (1987)
3. S. Noda, K. Tomoda, N. Yamamoto and A. Chutinan, *Science* **289**, 604 (2000)
4. S. Noda, A. Chutinan and M. Imada, *Nature* **407,** 608 (2000)
5. M. Imada, S. Noda, A. Chutinan, A, T. Tokuda, M. Murata and G. Sasaki, *Appl. Phys. Lett.* **75,** 316 (1999)
6. V. N. Astratov, V. N. Bogomolov, A. A. Kaplyanskii, A. V. Prokofiev, L. A. Samoilovich, S. M. Samoilovich and Y. A. Vlasov *NUOVO CIMENTO DELLA SOCIETA ITALIANA DI FISICA D-CONDENSED MATTER ATOMIC MOLECULAR AND CHEMICAL PHYSICS FLUIDS PLASMAS BIOPHYSICS,* **17**, 1349 (1995)
7. C. C. Cheng and A. Scherer, *J. Vac. Sci. and Technol. B*, **13,** 2696 (1995)
8. S. Y. Lin, J. G. Fleming, D. L. Hetherington, B. K. Smith, R. Biswas, K. M. Ho, M. M. Sigalas, W. Zubrzycki, S. R. Kurtz and J. Bur, *Nature*, **394**, 251 (1998)
9. S. Noda, N. Yamamoto and A. Sasaki, *Jpn. J. Appl. Phys.* **35**, L909 (1996)
10. N. Yamamoto and S. Noda, *Jpn. J. Appl. Phys.* **37**, 3334 (1998).
11. S. Noda, N. Yamamoto, M. Imada, H. Kobayashi and M. Okano, *J. Lightwave Technol.* **17**, 1948 (1999)
12. S. Noda, N. Yamamoto, H. Kobayashi, M. Okano and K. Tomoda, *Appl. Phys. Lett.* **75**, 905 (1999)
13. A. Chutinan and S. Noda, *Appl. Phys. Lett.* **75**, 3739 (1999)

Mat. Res. Soc. Symp. Proc. Vol. 681E © 2001 Materials Research Society

Multiple Wafer Bonding for MEMS Applications

M. Reiche [1], M. Haueis [2], J. Dual [2], C. Cavalloni [3], and R. Buser [4]

[1] Max-Planck-Institut für Mikrostrukturphysik, Weinberg 2, D – 06120 Halle, Germany
[2] ETH Zürich, Institute of Mechanics, Tannenstraße 3, CH – 8092 Zürich, Switzerland
[3] Kistler Instrumente AG, CH – 8408 Winterthur, Switzerland
[4] Interstate University of Applied Science Buchs, CH – 9471 Buchs, Switzerland

ABSTRACT

Most of the microelectromechanical systems (MEMS) require a 3-dimensional architecture which can efficiently be realized by multiple semiconductor wafer direct bonding. The present paper demonstrates the method on a force sensor for high resolution measurements of static loads. To minimize temperature stress an all-in silicon solution was developed in contrast to micromachined resonant force sensors published already in the literature.

The presented force sensor integrates load coupling, the excitation and detection of the vibration of the microresonator in one and the same single crystal silicon package. First measurements proved a sensitivity of 26 Hz/N and a resolution better than 3 mN.

INTRODUCTION

Different industrial applications require sensing devices which can measure a static load with very high resolution at normal and elevated temperatures up to 500°C. For a sensor having high stability single crystalline materials such as silicon, quartz, and galliumorthophosphate, respectively, are the obvious choice. Caused, however, by the existing manufacturing technology and availability, silicon is the preferred material.

In contrast to micromachined resonant force sensors, which have been published in the literature [1-3], we developed an all-in-single crystal silicon solution for the first level package where the load coupling, the vacuum encapsulation, the excitation, and detection of the vibration of the microresonator are integrated. To be operated in harsh environments the first-level packaged resonant structure is mounted into a robust steel housing, named the second-level package. The resonant structure is a bulk bending mode resonator. The present paper deals especially with preparation steps of the silicon structure (i.e. the first-level package) related to the semiconductor wafer direct bonding. Details of the general packaging concept and the sensor functionality are described elsewhere [4,5].

SENSOR DESIGN

The first-level package consists of a silicon structure prepared by multiple wafer bonding of 3 silicon wafers (figure 1). The first and second form a SOI (silicon on insulator) handle wafer including the resonator, the excitation, and the detection of the vibration. This structure is covered by a housing wafer forming a sandwich which has the resonator completely encapsulated. For the excitation a linear electrostatic force is generated by an integrated comb shaped electrode. A corresponding capacitive detection implies the problem of cross talk, which is solved by operating the oscillator in closed loop with a switched lock-in amplifier. The specific vibration detection layout made off chip electronics possible and allows a maximum degree of freedom for the design of the sensor chip [4]. In contrast to pressure sensors the coupling of the force to the resonator is the crucial issue of force sensors. A solution is proposed which guides the load from the metal housing via bolts to the silicon chip [4,5].

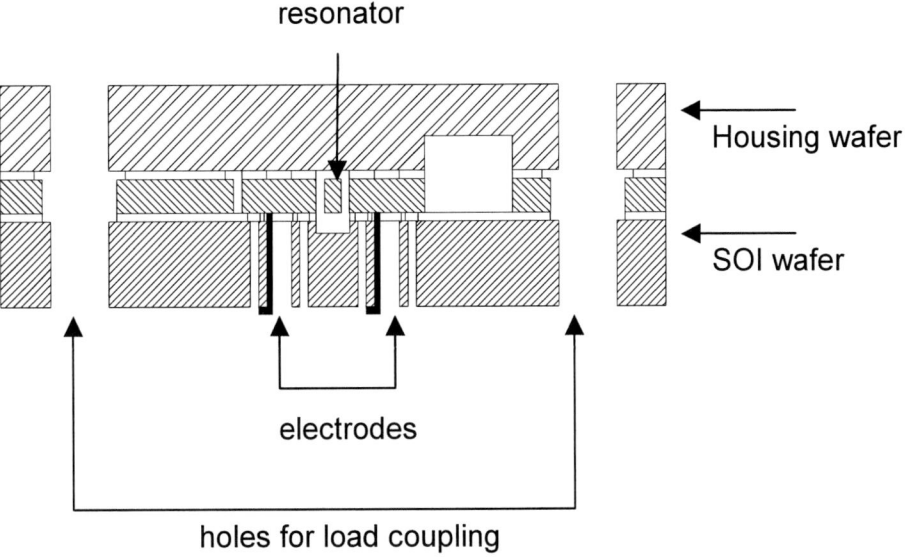

Figure 1. Schema of the silicon structure of the sensor.

The force translated from the bolts to the silicon chip is split into a part carried by the resonant structure and a part carried by the housing. Thus the maximum stress specifications of the resonant structure and the overload protection criterion can be matched. The facts that semiconductor wafer direct bonding is applied instead of gluing and silicon is the only material used minimizes the temperature stress and makes the system robust for high temperature applications.

Measurements of first sensors structures proved good load sensitivity, linearity, and repeatability. Figure 2 shows the resonance frequency of the sensor plotted versus the applied load. The initial frequency measured at 25°C amounts to 22654 Hz and a load sensitivity of about 26.4 Hz/N. The linearity error is negative and amounts to significantly less than ±0.5 %FSO calculated using the fixed point method.

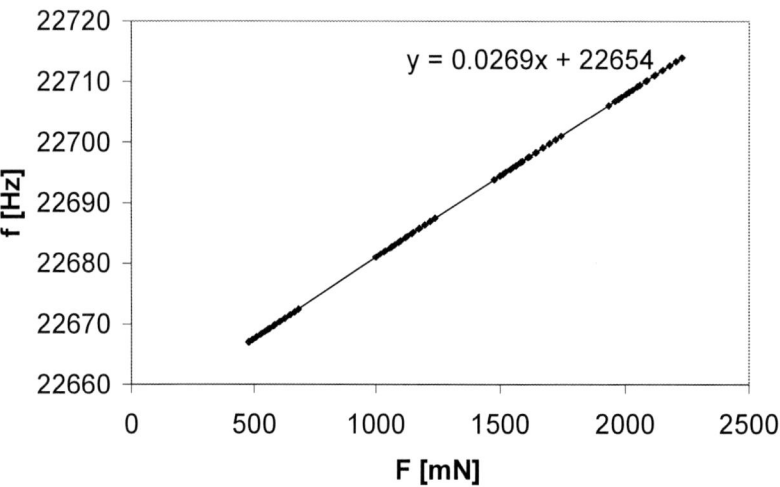

Figure 2. Resonance frequency of the force sensor plotted versus the applied load.

PREPARATION OF THE SILICON SENSOR STRUCTURE

The base of the sensor is a SOI wafer prepared from individual and prepatterned silicon wafers (diameter 4 in., (100)- orientation, p-type, $\rho = 0.01 - 0.02\ \Omega$cm). After oxidation both wafers are bonded at room temperature. In order to increase the bonding strength the wafer pairs are subsequently annealed at 1050°C (4 hours, O_2 ambient) or at 300°C (5 hours) with a preceding activation in an oxygen plasma. Both types of annealing result in a sufficient high bonding strength of more than 2 J/m². After bonding the wafer pairs were thinned by grinding and additional polishing (CMP process) resulting in SOI wafers having the following specification:
- Top layer thickness: 80 µm
- Buried oxide layer (BOX) thickness: 2 µm
- Handle wafer thickness: 350 µm

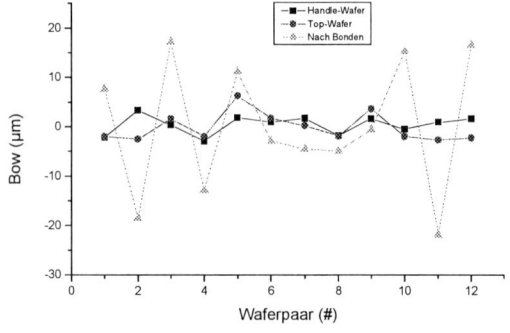

Figure 3. Bow of different SOI wafers (triangle). Squares and circles characterize the bow of the initial wafers used.

For the preparation of the SOI wafers a special preparation technique was applied resulting in low-stress wafers also by applying very thick BOX layers [6]. Figure 3 shows the bow of different SOI wafers as a measure of the stress. The typical bow of the SOI wafers is between ±20µm which is the same as for standard silicon wafers according to the SEMI standard. For further preparation steps (especially for the etching) it is important to deposit the oxide on the handle wafer.

The complete process sequence for the silicon sensor is shown schematically in figure 4. After preparing the SOI wafer (#1,2) the front and

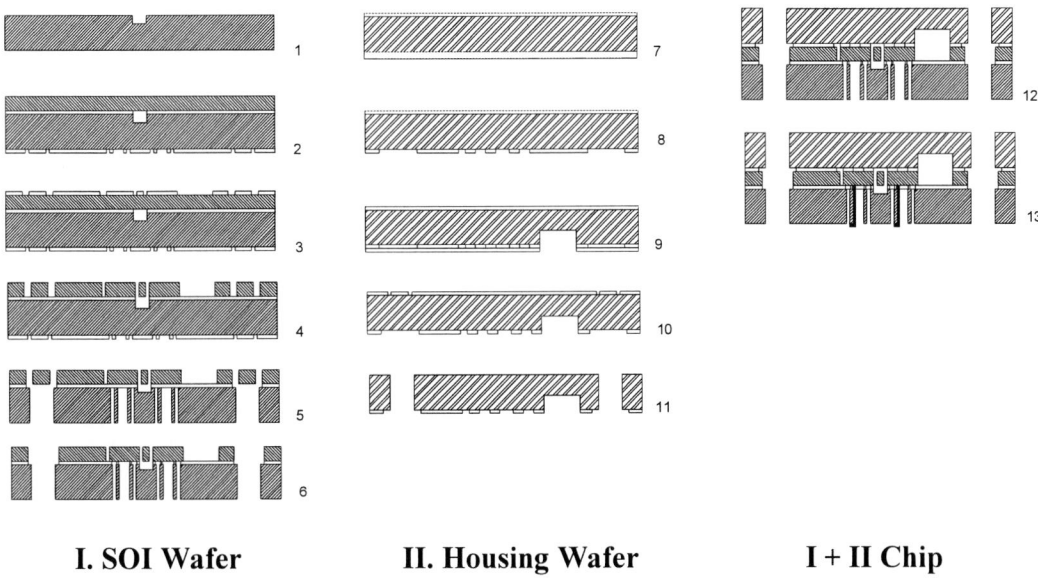

I. SOI Wafer   **II. Housing Wafer**   **I + II Chip**

Figure 4. Schematic presentation of the process sequence.

backside of the SOI wafer are patterned using a SiO_2 mask (#3) and etched by applying deep reactive ion etching (DRIE) with a SF_6 plasma (so-called Bosch process). Both, the handle wafer and top layer are completely etched (#5,6) in order to produce the sensor elements shown in figure 1. It is important to note that the mask oxide especially for the top layer is thick enough to prevent surface damages by the etching process (which can cause that the second wafer bonding is impossible).

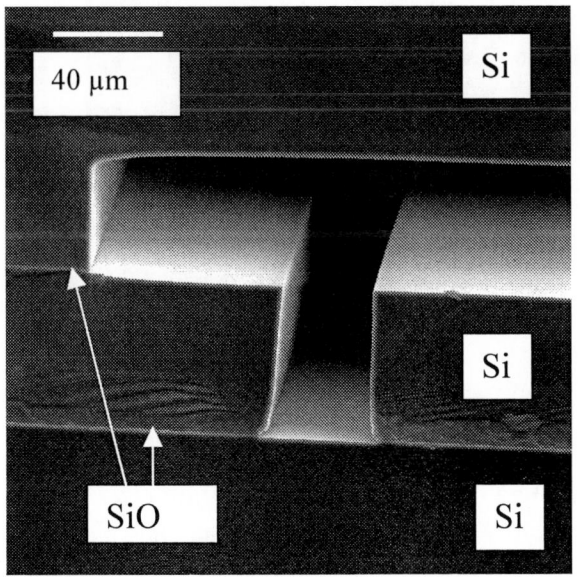

Furthermore, an additional housing wafer is patterned and etched using analogous DRIE processes (#7 – 11 in figure 4). After removing the mask oxides and deposition of a clean oxide layer on the surface of the housing wafer (thickness about 1.5 µm) the SOI and housing wafer are bonded by a second semiconductor wafer bonding (#12 in fig. 4). The wafer bonding was realized by a cleaning step with de-ionized water, an alignment of both wafers

Figure 5. Housing of the completely packaged silicon 3-wafer structure.
SEM cross-sectional image.

and the bonding under specific conditions. For the experiments the wafer bonding equipment of Karl Suess was applied (using the CL6 cleaner, the BA6 bond aligner, and the SB6 bonding equipment). The alignment error was about 10 µm. The bonding in the BA6 was carried out in vacuum ($p < 5 \cdot 10^{-4}$ Torr) followed by an annealing at 300°C for about 15 minutes (also in vacuum). This causes that the structure is vacuum sealed. After this procedure, the wafer stacks are subsequently annealed at 300°C or at 1000°C in order to increase the bonding strength. Figure 5 shows a cross-sectional image of the 3-layer stack.

The bonding behaviour and the bonding strength depend mainly on the surface quality of the patterned SOI and housing wafers. Wet chemical etching, for instance, causes an increasing surface roughness resulting in debonding. On the other hand, dry etching does not effect the surface roughness of both wafers. In general a spontaneous bonding is observed without any defects (bubbles). An infrared microscope image of the bonded wafer stack is shown in figure 6. The final step is the formation of the electrodes (#13 in figure 4).

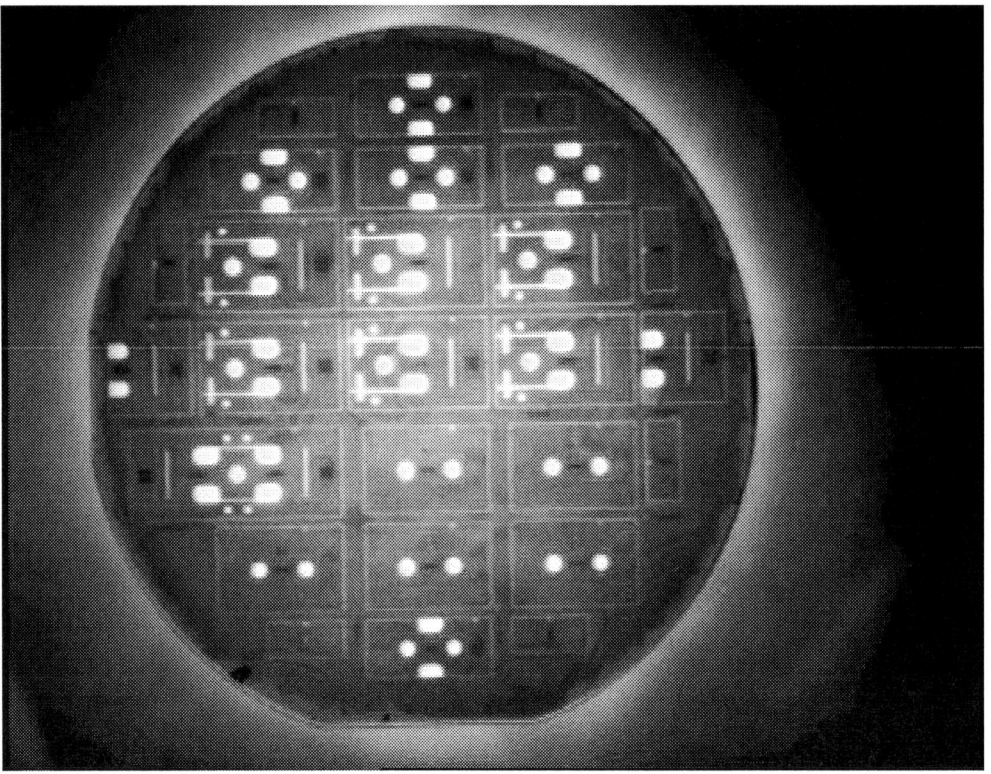

Figure 6. Infrared microscope image of the wafer stack produced by multiple wafer bonding.

CONCLUSIONS

Multiple semiconductor wafer direct bonding is applied to produce a packaged microresonator for measurements of forces in the order of 1N and below as required for industrial applications.

To minimize temperature stress an all-in silicon solution was developed in contrast to micromachined resonant force sensors published already in the literature. The following process steps are applied: (i.) Preparation of a SOI wafer package by a first wafer bonding step. A prepatterned and oxidized handle wafer was bonded to a top wafer. (ii.) The top wafer was thinned by grinding and polishing down to a thickness of 80 μm. (iii.) After patterning the top layer a third wafer (also patterned) was bonded on it. This second wafer bonding step includes a prealignment and bonding under vacuum followed by a low-temperature annealing.

The presented force sensor integrates load coupling, the excitation and detection of the vibration of the microresonator in one and the same single crystal silicon package. First measurements proved a sensitivity of 26 Hz/N and a resolution better than 3 mN.

REFERENCES

1. F.R. Blom, "Resonant silicon beam force sensor", Ph.D. Thesis, University of Twente, 1989
2. H. Tilmans, "Micro-mechanical sensors using encapsulated built-in resonant strain gauges", Ph. D. Thesis, University of Twente, 1993
3. R. Buser, N. de Rooij, IEEE Proceedings, 94 (1989)
4. M. Haueis, J. Dual, C. Cavalloni, M. Gnielka, and R. Buser in *Micro Systems Technologies 2001*, edited by H. Reichl (VDE Verlag, Berlin, Offenbach, 2001) pp. 261-266
5. M. Haueis, M. Gnielka, J. von Berg, J. Dual, C. Cavalloni, and R. Buser, *Transducers 2001* (2001) (in press)
6. U. Gösele and M. Reiche Recent *Developments in Semiconductor Wafer Bonding for SOI*, Proceedings of the Silicon-On-Insulator Conference, SEMICON Europa 2000, Munich, April 4, 2000

Mat. Res. Soc. Symp. Proc. Vol. 681E © 2001 Materials Research Society

Packaging Of Ultrathin Semiconductor Devices Through The ELO Packaging Process

Mike Sickmiller
ELO Technologies, Inc.
Torrance, CA 90501

ABSTRACT

The trend in semiconductor packaging is moving toward thinner and thinner packages. Likewise, chip profile is moving toward thinner and thinner chips. Presented here is a technique used to obtain a semiconductor package containing chips as slim as one micron in thickness. The ELO Packaging Process yields ultrathin chips for applications such as advanced heat sinking, high-efficiency optoelectronics, multiple stacked chips in a single package, and thin mechanically flexible semiconductor circuits. This technology is being developed around both the fab and packaging house so as not to interfere with the conventional semiconductor fabrication process flow.

Through a combination of back-grinding and chemical etch techniques, chips have been thinned to as little as 1 μm and bonded to a variety of new host substrates[1]. Several bonding methods have been utilized – including thin solder or epoxy layers – to bond these functional chips to a variety of new substrates. Ultrathin microwave power amplifiers have been bonded to heat sinks and optoelectronic devices have been bonded to transparent substrates. In both cases, the ultrathin chip configuration coupled with the desired substrate can increase performance of the chip by a factor of 10X.

INTRODUCTION

ELO Technologies, Inc. has devised a manufacturing process for packaging extremely high density, high power microelectronics through the use of epitaxial liftoff. Our ELO Packaging Process will improve electrical and thermal performance, increase circuit and package density, and increase system integration while reducing the size, weight, and cost of the package and system.

By removing the thermally and electrically insulating semiconductor substrate and coupling the electronics more intimately with a heat sink or CMOS processing circuitry, we can reduce the operating temperature and increase power density and interconnection speeds while minimizing electrical parasitics to the device and package[2]. The fundamental manufacturing technology has been proven and the effort is now focused on development of manufacturable prototypes including microwave power amplifier chips and VCSELs-CMOS integrated optoelectronic integrated circuits (OEICs). We expect performance enhancements which could reach an order of magnitude increase in power density, reliability, efficiency, speed, or any combination of these parameters.

MOTIVATIONS

This technology will enable the next generation of high-speed, high-power, high thermal efficiency wireless and Opto-electronic integrated circuit semiconductor components. The ELO Packaging Process allows for the thinning of semiconductor chips to only a few microns in thickness and their bonding and electrical and thermal interconnection to virtually any substrate. Thermal power dissipation has been identified as the largest hurdle in GaAs microwave devices and in many cases is the limiting factor in die shrinkage [3, 4]. As a result of the technology presented here, an ultrathin microwave power amplifier can be integrated directly onto a heat sink for improved operating efficiency resulting in higher power output, or GaAs VCSELs can be integrated directly onto CMOS circuitry for a high-speed OEIC.

Applying this process to VCSELs and providing backside interconnections, direct placement onto a silicon CMOS chip is possible. Placing a small VCSEL array onto a chip and attaching a fiber bundle will allow high-speed multi-channel communications in the Terabit per second range[5]. This is very promising for low cost and high performance optical data links. Using optical interconnects will also reduce electromagnetic interference (EMI) and heat production associated with conventional metal interconnects and wire bonds.

Hybrid Opto-electronic integrated circuits (OEICs) are promising for many cost-effective applications such as vertical optical interconnection for high data rate communication. Vertical optical interconnection through OEIC techniques will reduce in-plane interconnections and input/output bottlenecks, thereby increasing overall processing speed and reducing packaging dimensions significantly. Optical interconnects also promise high bandwidth, low capacitance and low crosstalk.

PROCESS OVERVIEW

The ELO packaging process begins with an etch to define the individual devices. During the substrate removal process, the separate die will be formed.

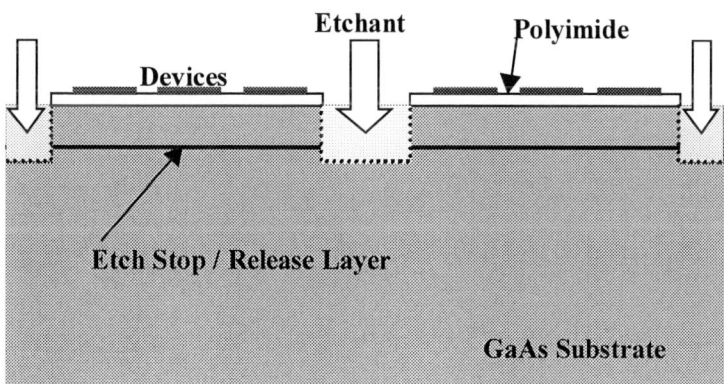

Figure 1. Etch to define devices or circuits on wafer.

The next step of the process is to attach the intermediate carrier. The function of the intermediate carrier is to provide mechanical support for the ultra-thin devices and to act as a handle for subsequent processing steps.

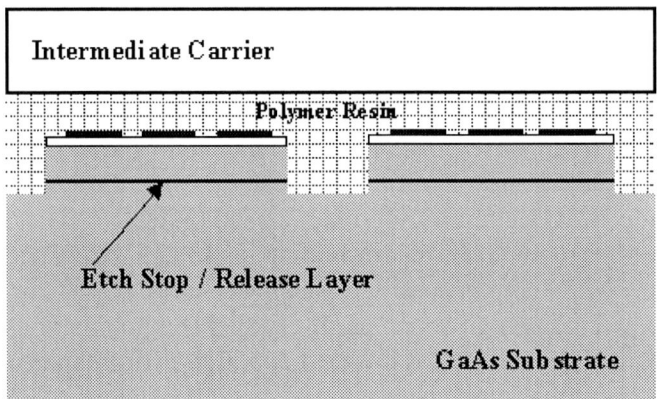

Figure 2. Bonding of intermediate carrier to wafer.

Once firmly attached, the semiconductor substrate is etched back to a final thickness of about 3-5μm, determined by the location of the etch stop layer.

Backside processing follows, using traditional photomasking methods. This step is necessary to accurately position the metal solder pads under heat generating elements in the circuit, such as transistors. Various methods can be used such as sputtering, electron-beam evaporation, or electroplating.

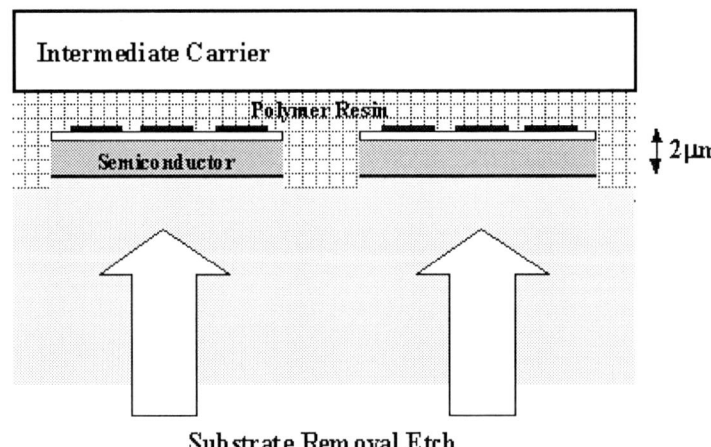

Figure 3. Etch processing to remove substrate.

142

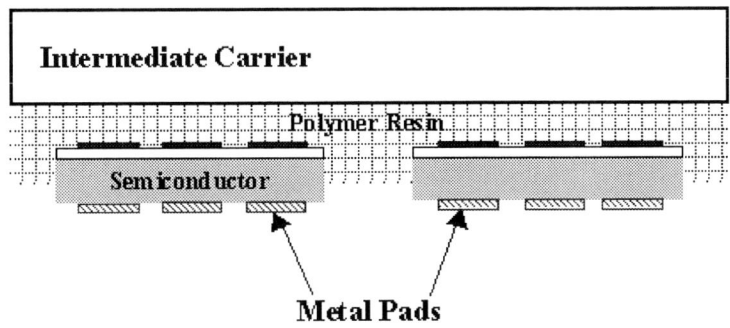

Figure 4. Backside processing - metallization of the ultra-thin film semiconductor.

With the metals deposited, the ultra-thin devices can be solder-bonded to the heat sink. Because the thin film is supported by the Intermediate carrier, this step can be done with a conventional Flip Chip Bonder.

After the solder bonding step (epoxy can also be used as the bonding material), the intermediate carrier is then removed using a solvent to dissolve the adhesive. The key to this step is the materials compatibility of the temporary adhesive used to bond the thin devices to the temporary carrier. The adhesive must be robust enough to withstand process temperatures, but must be able to be removed in a gentle enough manner such as not to damage the devices or substrate materials during the process.

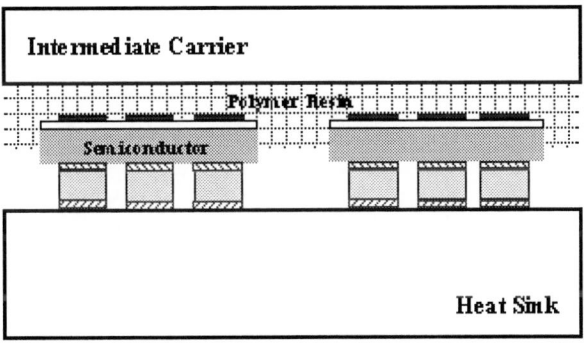

Figure 5. Solder bonded devices with carrier attached.

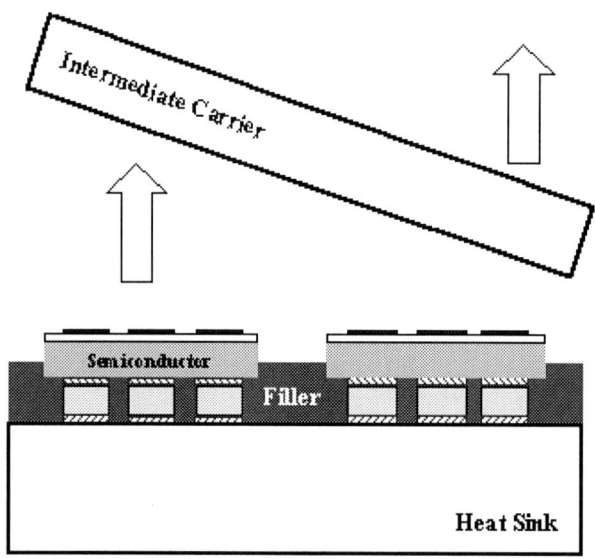

Figure 6. Intermediate carrier removal done by dissolving the polymer resin.

The resulting product is an ultra-thin GaAs device mounted upon a heat sink with solder heat pipes intimately coupling the chip's hot spots to the sink. Essentially, the GaAs substrate has been replaced with a highly thermally conductive substrate yielding increased thermal performance with no expected loss in electrical performance.

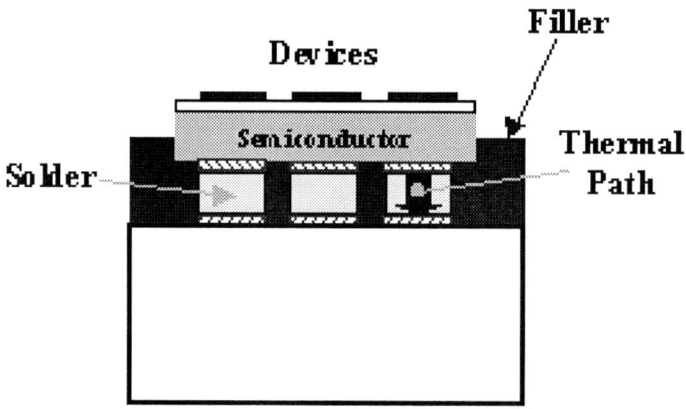

Figure 7. Final product of ELO packaging process.

CONCLUSION

The ELO Packaging Process can be used to achieve the next generation in chip packaging – that of ultra-thin devices in packages and chip-to-chip interconnections. The applications include thinned VCSELs interconnected to silicon CMOS chips where terabit optical interconnection speeds will be possible and high power microwave amplifiers thinned to only a few microns eliminating the thermal resistance of the GaAs substrate. This technology will provide a low cost, high yield solution to device integration with any arbitrary substrate that can be added into any manufacturing line with minimal process disruption.

REFERENCES

[1] A. Pike, S. Jacobs, M. Sickmiller, "Novel Packaging for Superior Thermal Performance in High Power, High Frequency Devices," IMAPS SoCal Conference 1999 (1999).

[2] M. Sickmiller, A. Pike, L. Vo, S. Jacobs, E. Yablonovitch, "Ultra-High Efficiency Light Emitting Diodes Through Epitaxial Lift-off Packaging," Conference Proceedings of IMAPS International Symposium on Microelectronics 114 (Nov. 1998).

[3] K. Yamamoto, et al "A 3.2-V Operation Single-Chip AlGaAs/GaAs HBT MMIC Power Amplifier for GSM900/1800 Dual-Band Applications", IEEE MTT-S Digest, 1999, pp1397-1400

[4] J. Griffiths, V. Sadhir, "A Low Cost 3.6V Single Supply GaAs Power Amplifier ID for the 1.9GHz DECT System", IEEE Microwave Systems Conference, 1995, pp37-40

[5] J. Tatum, "Packaging flexibility propels VCSELs beyond telecommunications," Laser Focus World April, 131-136 (2000).

Mat. Res. Soc. Symp. Proc. Vol. 681E © 2001 Materials Research Society

A Novel Ultra-miniature catheter tip pressure sensor fabricated using silicon and glass thinning techniques

Henry Allen[1], Kamrul Ramzan[1], Jim Knutti[1], and Stan Withers[2]

[1]Silicon Microstructures Incorporated, 46583 Fremont Boulevard, Fremont, CA 94538
[2]Jomed Inc, 2870 Kilgore Road, Rancho Cordova, CA 95670

Abstract: A novel subminiature pressure sensor for blood pressure measurement has been fabricated. The device is only 250 microns wide and 70 microns thick. It is 1.1 mm in length. The sensor is housed in a guide-wire lead for use in measuring coronary artery blood pressure. The device has a 5 micron thick silicon diaphragm and senses pressure using a 1/2 bridge piezoresistive network. Glass is processed to provide depressions above the sensing area as well as above the connection area of the device. A full-thickness silicon wafer is processed using standard micromachining techniques. V-Groove notches are micro-machined on the top surface of the silicon to provide locators/guides for the lead-wires. Diaphragm windows are patterned on the back of the silicon wafer and the wafer is etched down to form the 5 micron diaphragm, using electro-chemical etch-stop techniques. The Glass and Silicon wafers are aggressively cleaned prior to bond. The glass and silicon wafers are then precisely aligned to better than 10 microns and bonded using anodic bonding techniques.

The glass/silicon wafer sandwich then has the silicon thinned from 400 microns to 37 microns using both grinding and polishing. Then the full-thickness glass wafer is etched in HF to a thickness of 37 microns as well, for a composite 74-micron thick structure. The wafer is then diced to form the micro-mechanical structure.

INTRODUCTION

One of the challenges of invasive monitoring of the body is in making parts small enough to allow them to fit into the targeted environment. One such application is in the small arteries of the heart. Here the arteries start at 2 to 3 mm external diameter and rapidly bifurcate into progressively smaller arteries in the sub-millimeter range. A catheter that occupies more than 1/2 the area of the vessel under study will have detrimental impact on accuracy of the measurement. Because of these concerns, a pressure sensor has been fabricated in a guide-wire to allow accurate measurements of pressure in the Coronary arteries. The guide wire is 0.355 mm in diameter and requires a sensing element that can be housed with lead-outs, housing, and encapsulation in that diameter.

The completed guide-wire is shown in Figure 1. The diaphragm size is 280 microns X 130 microns. The wire is constructed with a helical sheath for flexibility except at the tip and around the sensor itself. Here a stainless-steal housing with a machined-in lumen for the sensor is used. Three 25 micron leads are connected to the sensor to allow measurements from 2 piezoresistive-sensing elements. A simplified cross-section of top-view of the device is shown in Figure 2.

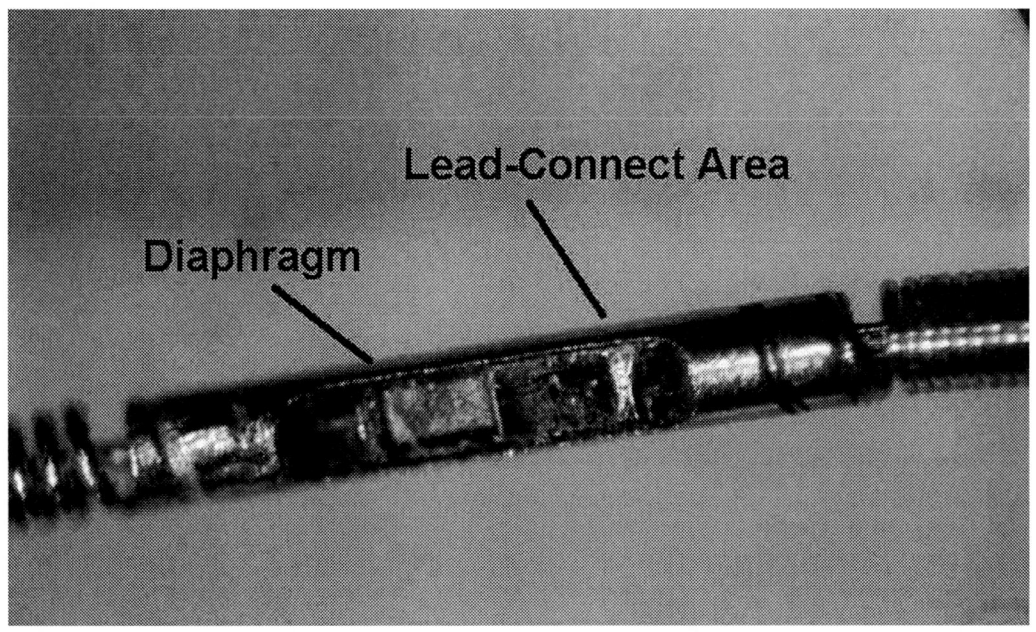

Figure 1. Completed SmartWire Guidewire Pressure Transducer

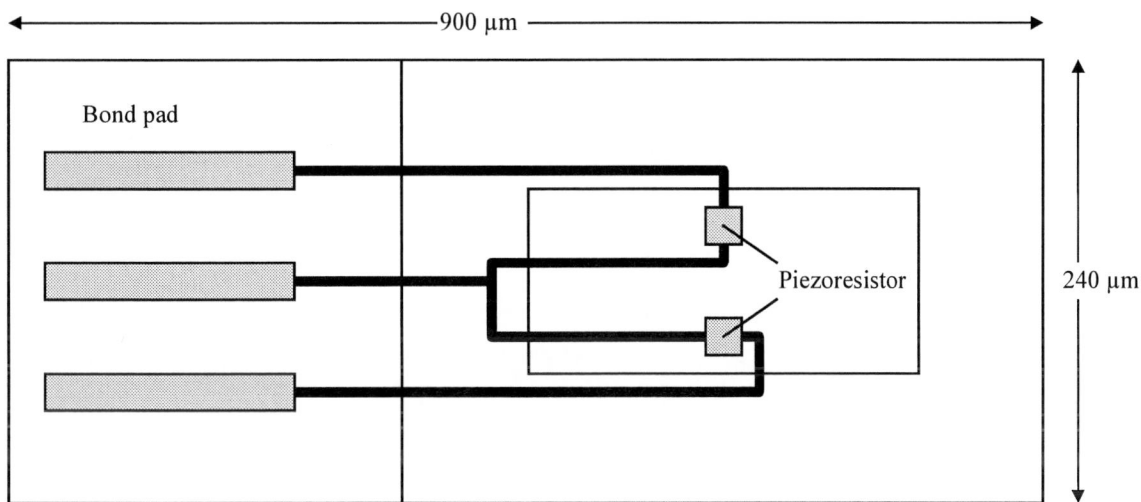

Figure 2. Schematic diagram of sensor showing external dimensions

DESIGN

A number of design constraints were established at the outset. These included having the lead-outs and the sensing surface on opposite sides of the die, having a sealed cavity to provide a reference pressure incorporated on chip, and having the pads designed for easy alignment and attachment of the leads. The chosen configuration is shown schematically in Figure 3. In this case, V-grooves on the top-side of the wafer provide for lead-out alignment. Further, because of pad design, the wires can successfully be recessed by 30 to 50% of their actual height.

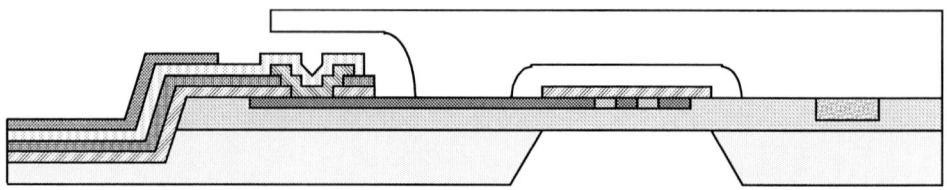

Figure 3. Cross-section of the Pressure sensor. Leads are connected in the trench area to the left of the figure. The top glass is etched to form a sealed cavity as a reference pressure.

The cap is a glass cap. This choice was made to allow for a simple metalization because the glass can be anodically bonded to the silicon wafer. It also allows two steps in the glass, one over the diaphragm area and one over the pad area. Further it allows easy visualization of the circuit area prior to dicing the wafer and facilitates easy alignment of glass to silicon prior to bonding.

The active sensing area is formed with standard KOH etching techniques. The main variant on this is that the wafer is thinned down after the glass is bonded to the silicon so that the apparent sidewall edge is radically reduced. The actual cross-section of the sensor and die is shown in Figure 4.

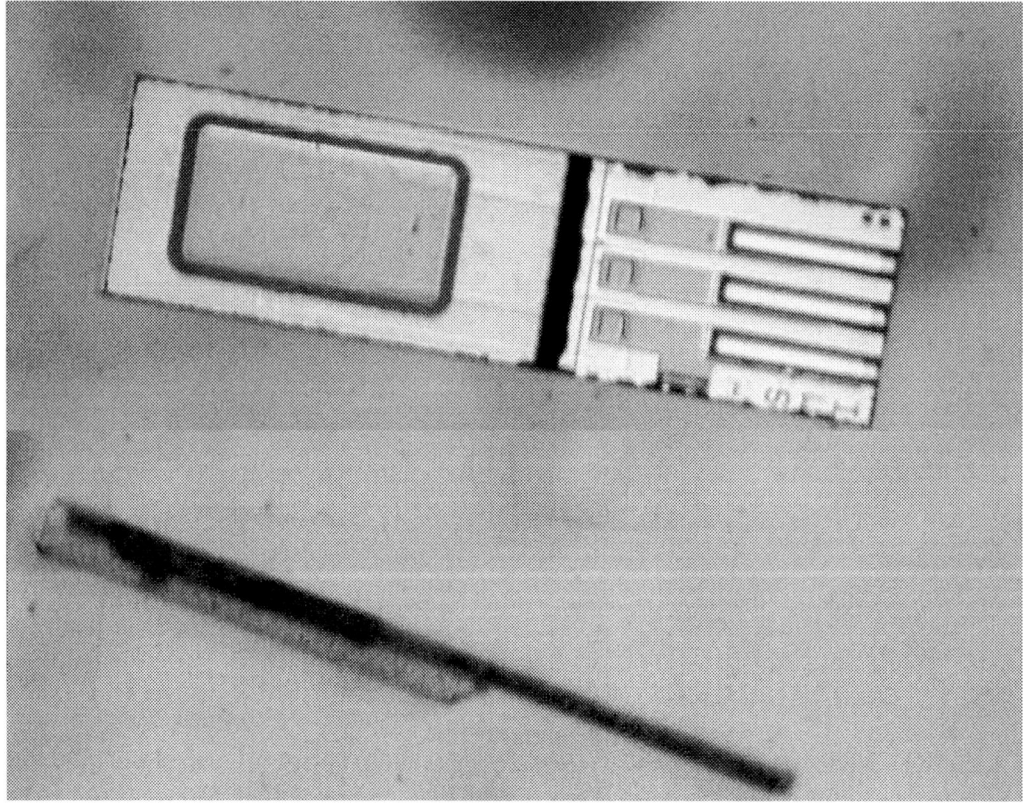

Figure 4. Diced sensor looking at the top of the silicon and the side-view of the sensor showing the glass cap (bottom left) and the thin silicon as diced.

Resistor Layout Considerations Because of the need for a very narrow structure, analysis of the optimum geometry and layout for resistors was preformed using conventional Finite-Element modeling software. The rectangular diaphragm shape means that there is no stiffening of the narrow dimension of the sensor by the perpendicular dimension, unlike the conventional square diaphragm. Peak stress of equal and opposite magnitudes are observed at the center and edge of the rectangular diaphragm. However for ease of routing the resistors, the chosen location of the resistor pair is on the outside edge of the diaphragm, using tangential and perpendicular resistors.

PROCESS SEQUENCE

The sensor is fabricated on 400 micron thick p-type <100> starting material with a thin epitaxial layer. N+ regions to provide contact to the substrate are diffused into the Epi. P+ and P-layers are Implanted and then a KOH etch is used to form the V-groove recesses in the pad area. The wafer is then oxidized and contact is made thru the oxide to the P+ regions. The Wafer flow thru the foundry part of the operation is shown in Figure 5. The wafers are then metalized, a pattern is aligned for the back cavities and the wafers are then electro-chemically etched. The diaphragm thickness is determined by the epitaxial thickness. Wafers then undergo a second metalization of PtSi/Ti/W/Au and the Pad metal is defined into the V-groove areas. Wafers are then probed and the sensitivity of the wafers are at select points. A final top-surface lithography and etch are done to define the anodic bond surface. The silicon wafers are processed thru a sulfuric-peroxide clean and an HF dip as the final clean prior to bond.

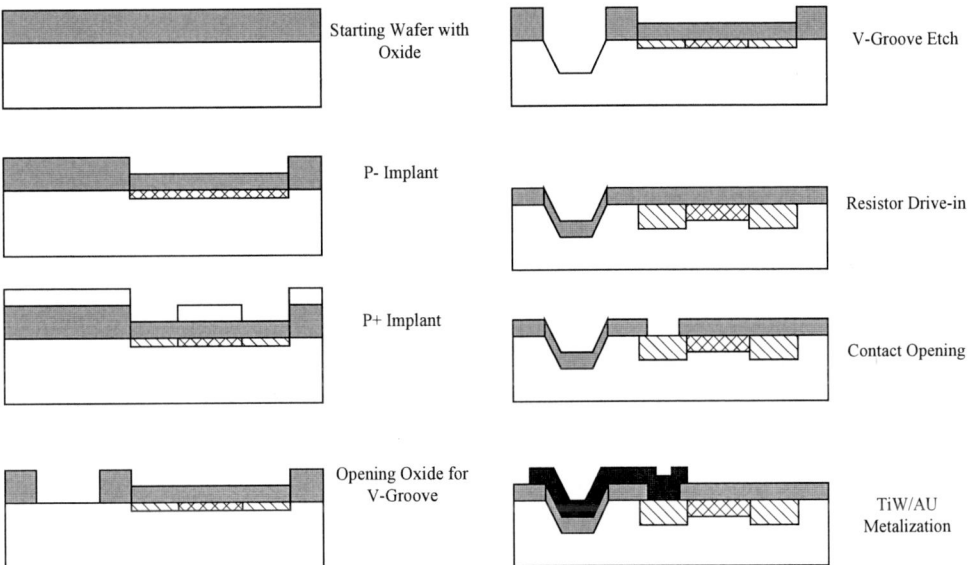

Figure 5. Top-side Sensor process

Concurrent with the silicon processing, glass is also processed. As mentioned previously, the glass has two different depths - one for the area over the sensing area and a deeper one over the area used for the lead-outs. The 500-micron glass wafers are initially metalized with CrAu. The deeper area is first defined and the CrAu removed in these areas. The glass is then etched in BOE. This process is then repeated for the shallower area. The depth of the deeper area is then the sum of the initial etch depth plus the second etch depth. The metal is stripped after the shallow etch and the wafers cleaned prior to anodic bond.

The glass and silicon wafers are aligned and the wafers are then anodically bonded. The next composite sandwich at this point is 900 microns (400 micron Silicon and 500 micron glass). Wafers are ground and polished on the silicon surface to result in a 37 micron nominal thickness. The wafers at this point are 537 micron thick (37 micron silicon and 500 micron glass). Wafers are then placed in straight HF to thin the glass down to a nominal 37 micron thickness to produce a composite thickness of 75 microns. The thining sequence is shown in Figure 6. Table 1 shows typical thickness control on both the glass and silicon.

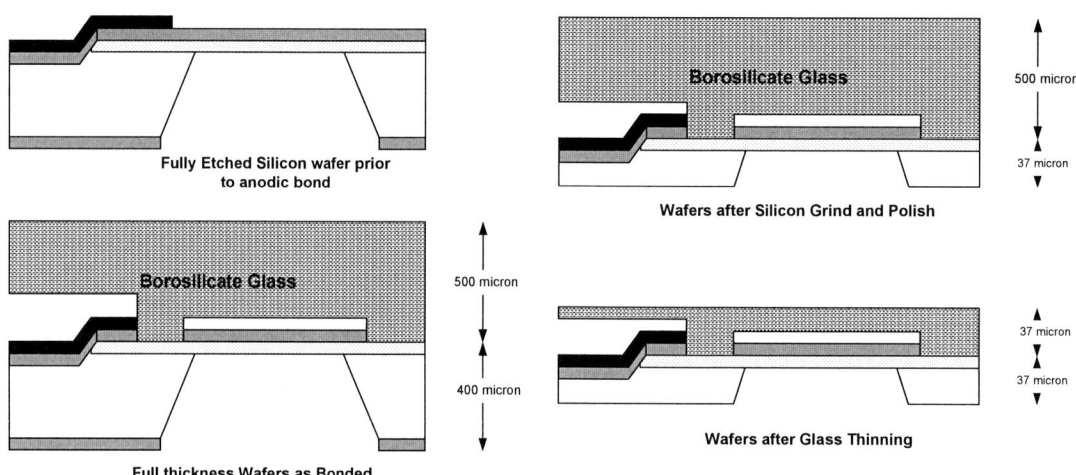

Figure 6. Bonding and thinning sequence

Table 1. Thickness Control (in microns) for 3 wafers from one lot of wafer

Position	Post grind & etch-Wafer A			Post grind & etch-Wafer B			Post grind & etch-Wafer C		
	Silicon	Glass	Total	Silicon	Glass	Total	Silicon	Glass	Total
1	35	40	75	38	31	69	37	31	68
2	37	38	75	38	36	74	37	36	73
3	36	39	75	37	37	74	36	33	69
4	38	37	75	35	35	70	37	43	80
5	38	29	67	35	31	66	36	38	74
6	37	36	73	35	34	69	36	44	80
7	36	42	78	36	37	73	37	38	75
8	36	41	77	36	34	70	37	40	77
Average	37	38	74	36	34	71	37	38	75
Std. Dev.	1.1	4.1	3.3	1.3	2.4	2.8	0.5	4.5	4.5
Min.	35	29	67	35	31	66	36	31	68
Max.	38	42	78	38	37	74	37	44	80

The composite structure is very fragile because, unlike thin silicon which can bend, the glass-silicon structure has two different material characteristics and fractures in the glass tend to propagate thru both layers resulting in broken wafers.

Figure 6 shows the view of the top silicon surface looking thru the glass. A close-up of one die is shown as an insert. The three gold fingers are the electrical contact area. The "tick" marks at the left and right of the die (as well as at the top and bottom) indicate saw alignment targets for dicing the die. The die is made in a relatively sparse array so that the full thickness wafer can be etched before polishing. This requires a reduction in density. However, for the application, the reduced density is not as important as the ease of manufacturing.

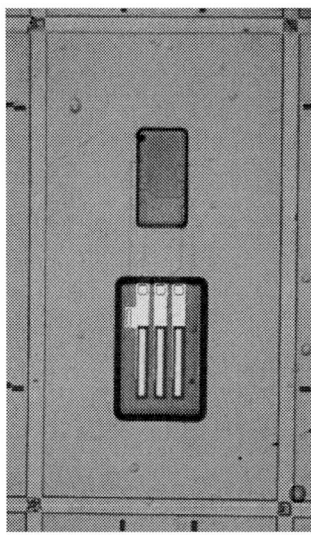

Figure 6. Wafer level view of 75-micron thick glass and silicon sandwich. Figure to the right is a picture of a single die from the wafer, looking thru the glass down onto the top silicon.

CONCLUSIONS

A robust process has been developed for the manufacturing of an ultra-miniature pressure sensor for catheter applications. The process uses anodic bonding with precision alignment requirements, precision silicon grinding and polishing, and glass thinning. The resultant device is only 240 microns or less when sawn by 900 microns long by 74 microns thick. The composite sandwich is composed of about 37 microns of silicon and 37 microns of glass. The process is such that full 100-mm wafers can be delivered for dicing on a routine basis.

ACKNOWLEDGEMENTS

The authors would like to acknowledge Canh Pham for expertise in the anodic bond and alignment of the glass and silicon sandwiches. In addition, the product is only feasible due to the precision dicing and lead-attach expertise at Jomed. Special thanks to Steve Rodriguez at SMI for scheduling the production runs, to Chris Ingman and Mike Meyer at Jomed for manufacturing expertise, and to Mike Eberle at Jomed for continued guidance and support on this project.

Mat. Res. Soc. Symp. Proc. Vol. 681E © 2001 Materials Research Society

Bonding, Splitting and Thinning by Porous Si in ELTRAN® ; SOI-Epi Wafer™

Kenji Yamagata and Takao Yonehara

ELTRAN Business Center, Canon Inc.

6770 Tamura, Hiratsuka-city, Kanagawa 254-0013, Japan

ABSTRACT

ELTRAN is a unique technique to produce the SOI wafers using a porous Si material in semiconductor process. In ELTRAN process, it is required to form the porous Si layer on entire wafer surface uniformly, stably and mass productively without contaminations. In this investigation, we have carried out the simulation of current density distribution to unify the porous layer thickness by finite element method. Canon designed and completed an automatic anodization apparatus. As a result, we could produce the 8 and 6 inches porous Si wafers and ELTRAN SOI wafers stably. And we also developed successfully 300mm ELTRAN SOI wafers with excellent SOI film thickness uniformity.

INTRODUCTION

The porous Si material has been investigated in some peculiar fields due to its unique features. For example in semiconductor field, FIPOS (fully isolation by porous oxidized silicon) was reported at about 20 years ago (1). However, it was not produced mass productively. About 10 years ago it was found that the porous Si emit visible light (2). In 1990 Canon started to develop ELTRAN® SOI-epi wafers using porous Si. The results were reported in 1994 (3)(4).

ELTRAN is one of the SOI wafer technologies categorized as bonding and etching SOI wafers. Its production method is shown in figure 1 and explained below.

The starting material is a P+(0.01-0.02 ohm-cm) Si wafer that is called "seed wafer". In the first step, the seed wafer is immersed in the HF solution and applied electric current to form porous structure on the wafer surface. And it is also possible to make the multiple layer structure with different porosity by changing applied current density. Scanning electron microscopy (SEM) image of porous Si is shown in figure 2. In our case, doubly layered porous Si is formed.

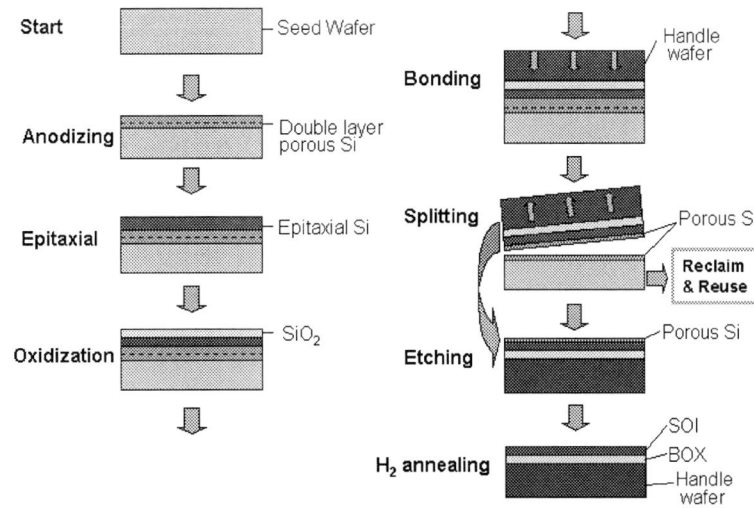

Figure 1. ELTRAN process flow.

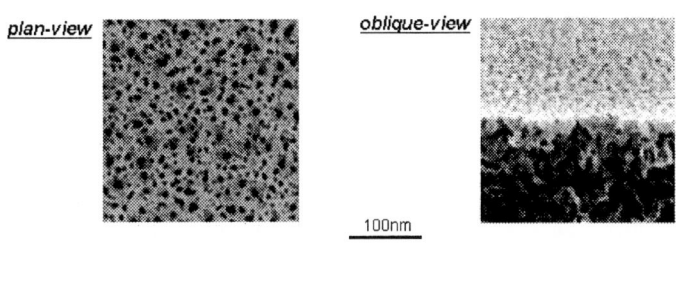

Pore size 6~8 nm, pitch 10~30 nm
Pore density ~ 10^{11} cm^{-2}, Porosity ~20%

Figure 2. SEM images of porous Si surface. The left one shows the plan-view and right one is the oblique-view.

Next, epitaxial layer (SOI layer) is grown on the porous layer. And the buried oxide (BOX) layer is formed by oxidization the epitaxial layer surface. Next, "a handle wafer" is prepared. After cleaning both of the seed wafer and handle wafer, they are bonded with each other. Then the bonded wafers are annealed at 1100 degrees C in order to enhance the bonding strength. By this treatment, the bonded 2 wafers are unified due to covalent bond. The bonded wafer is split open by a water jet machine at the interface of the 1st and the 2nd porous layer (5). It was found that the stress is concentrated at that interface of the porous layers, and the water jet triggers to release the stress resulting to separate the bonded pair from the interface. After the splitting, porous Si layers remains on both of the wafer surfaces. In the next step, the porous layer on the handle wafer is etched off selectively. The etching selectivity of porous Si against epitaxial Si is about 10E5, so it is possible to etch the porous Si without degrading the SOI thickness uniformity. The porous Si on the seed wafer is also removed. Then the seed wafer is recycled as a starting material. We have already confirmed the

153

reliability of recycling as many as 4 times at the same quality. The last step of ELTRAN process is hydrogen annealing. The wafer surface just after porous Si etching is as rough about 10nm in Rrms value (10*10um area). But when it is annealed in hydrogen ambient at high temperature over 1000 degrees C, the surface becomes atomically smooth due to Si migration phenomenon.

In ELTRAN process, porous Si material plays an important roll. The requirements for the porous Si are as follows,
1. To obtain the high quality epitaxial layer, the pore density should be controlled and formed on the entire wafer surface without contamination.
2. To split completely at the inside of porous layer, the stress controlled doubly layered porous Si is formed within +/- 10% thickness uniformity.
3. To etch the remaining porous layer selectively after splitting, the porous structure (porosity) is optimized.
We have developed a fully automated anodization system in order to satisfy those requirements.

EXPERIMENTS

(a) Design of anodization apparatus
The most important requirement to porous Si in ELTRAN process is that the porous layer should be formed on entire wafer surface. For the purpose, we developed the vacuum chuck holder that holds the wafer at the peripheral area of the back surface. And we have designed the apparatus that can handle multiple wafers in a batch. The structure is shown in figure 3. In this figure, the only 2 wafers are drown, but the actual apparatus can handle several tens or several hundred wafers in theory. In this time, it was optimized the only wafer-1 and wafer-2 because between wafer-2 and some wafers set at behind wafer-2 will become same features. We have decided to be 5 wafers batch apparatus.

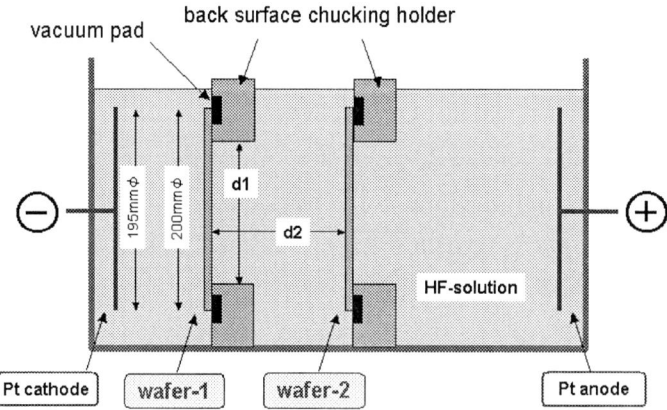

Figure 3. Illustrations of novel anodization apparatus for 8 inches wafers (basic design)

(b) Simulation of current density in anodization

Among anodization process parameters, the porous Si thickness uniformity is the most important thing especially for the water jet splitting to recycle the seed wafers. In here, we focus the optimization of porous thickness uniformity.

In experiments and simulations, PTFE material was used for the anodization bath and the holder. The electrode material is Platinum. The electrolyte is mixture of 49% HF : ethanol = 2 : 1. Si wafer is P type, 0.01-0.02ohm-cm, 200mm in diameter. Both the actual experiment and simulation were carried out. Main parameters are "d1"(holder orifice size) and "d2"(wafer to wafer distance) in figure 3, and they are variable.

The three dimensional current density distributions were simulated by finite element method. Calculated points were meshed in allover the electrolyte of radius area of the Si wafer because the current density becomes symmetry due to the circular orifice design of holder structure. Especially the calculated points were meshed finely at a changing point of the wafer or wafer holder. The result is shown in figure 4.

Amplitude of calculated current density is shown as the length of the arrow (but it is too small to see the length). In the data of calculation, the variation of the current density from the wafer center to the edge is not so large in wafer-1. On the other hand in wafer-2, the current density at the wafer edge is extremely small.

In order to evaluate the current density distribution quantitatively, we tried to plot the current density values on the virtual plane exactly 10um above the wafer surface. It is shown in figure 5 with the experimental data.

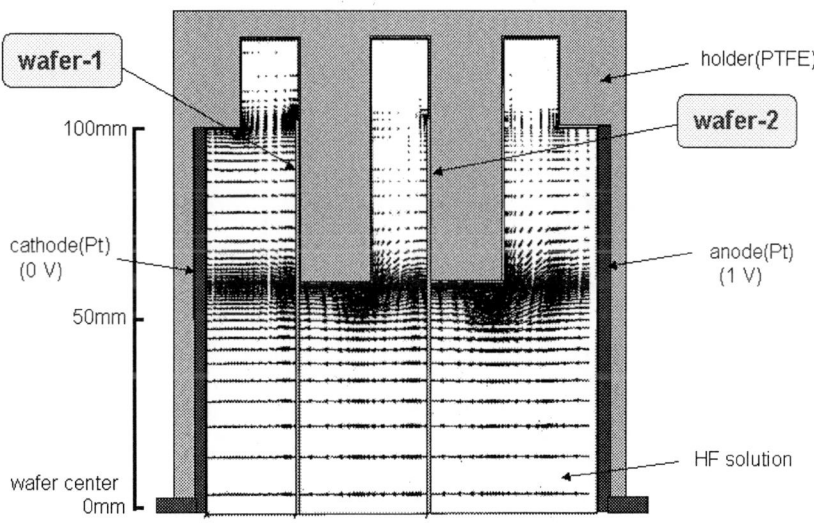

Figure 4.   The image of current density simulation.

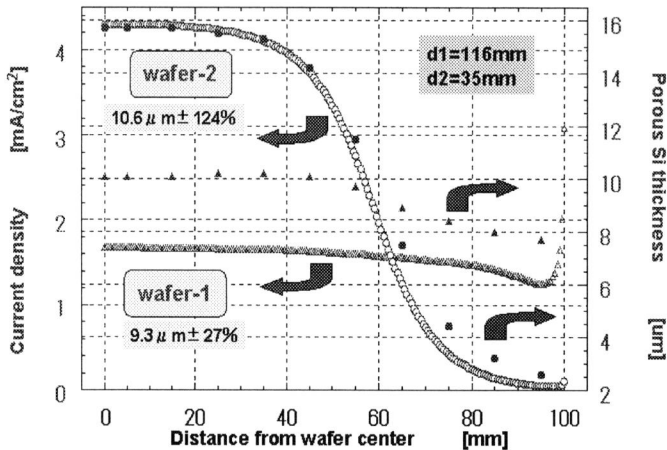

Figure 5. Simulated current density distribution and actual porous thickness data.

DISCUSSION

(a) The improvement of uniformity in porous layer thickness

Figure 5 indicates that porous thickness uniformity of the wafer-1 is relatively good. The wafer-1 is facing to the same size of cathode. But for the wafer-2, the cathode is the back surface of the wafar-1 covered with the orifice of the diameter"d1". Accordingly in the wafer-2, the porous Si thickness near the wafer center is big and in the peripheral area it is extremely small. Same result is expected for the wafers that would be placed at the anode side from the wafer-2.

Then we have tried to expand the orifice diameter from 116mm to 172mm, calculated and carried out experiments again. That result is shown in figure 6.

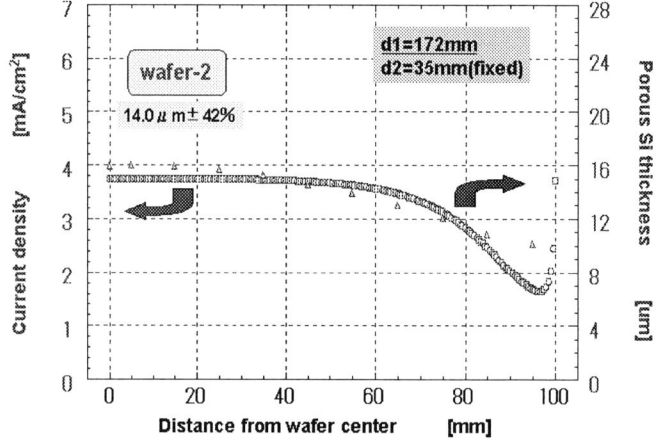

Figure 6. Improvement of "d1" orifice size.

It was found that it is possible to improve the porous thickness uniformity by expanding the orifice diameter. But the variation is still 42% (experimental data), it is not enough optimized.

Next, we tried to expand the distance between the wafer-1 and the wafer-2 from 35mm to 80mm at maximum. That result is shown in figure 7. The solid red line is the simulation data, and the dotted blue line is the average of the experimental data. The vertical axis is the thickness range of the porous Si layer. The values are calculated as (max. − min.)/average thickness. By this graph, we found that the thickness range of +/-10% (range:20%) is achieved when the wafer to wafer distance is larger than 80mm. It was found that the more it takes distance of wafer to wafer, the better the thickness distribution. But we decided the distance as 65mm considering the apparatus dimension.

We made a fully automated anodization apparatus with the 5 wafers batch system referring the previous result, installed. Using this system, we have carried out the anodization of 5 wafers batch, and obtained thickness uniformity data shown in figure 8.

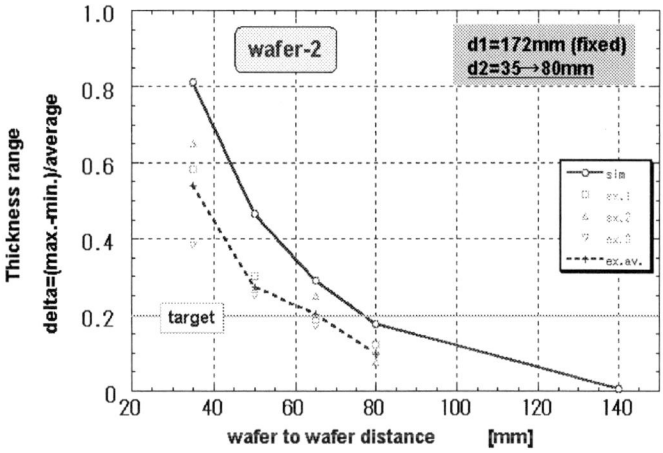

Figure 7. Improvement of "d2" wafer to wafer distance.

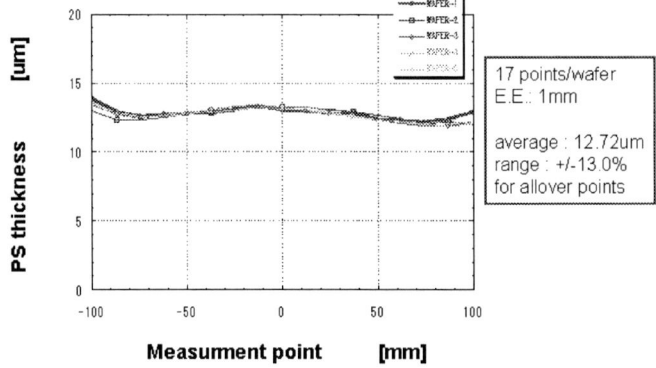

Figure 8. The porous thickness uniformity of 5 wafers in a batch.

In figure 8, the red line shows the wafer-1, and the other blue lines show the wafer-2 to the wafer-5. This result indicates that the all of 5 wafers have same distribution and it is possible to control the porous Si thickness uniformity as 12um +/- 13% with 9 points on the wafer surface 1mm inside the wafer edge.

(b) Production stability in ELTRAN

We also developed an automatic bonding machine, a water jet splitting machine and a porous Si etching machine for ELTRAN process. Currently we are producing 6 and 8 inches SOI epi-wafer mass productively. Its production stability is indicated in figure 9 and figure 10. Figure 9 shows the SOI thickness stability and the particle evaluation of the sample of 6 inches ELTRAN wafers of which the SOI and the BOX thickness are 30nm and 100nm respectively. Figure 10 also shows the same evaluations of 8 inches ELTRAN wafers, but the SOI and the BOX thickness are 200nm and 200nm.

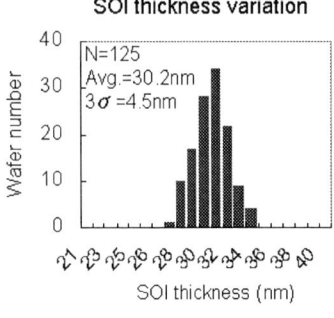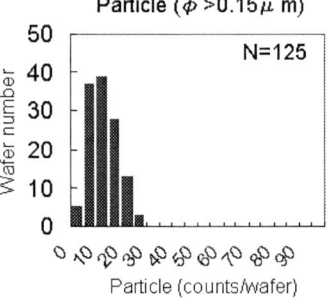

Figure 10. Production stability: example-1 (6 inches ELTRAN SOI/BOX=30nm/100nm)

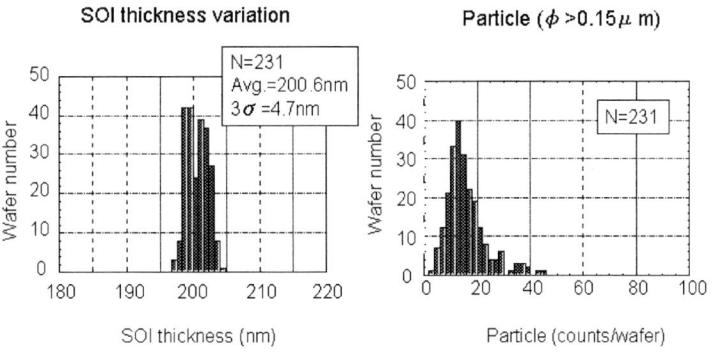

Figure 11. Production stability: example-2
(8 inches ELTRAN SOI/BOX=200nm/200nm)

The results of 6 inches SOI wafer thickness variation and those of 8 inches SOI wafers are nearly the same as 4.5nm and 4.7nm in 3 sigma, and the particle level is almost same.

(c) 300mm ELTRAN

We already succeeded to produce 300mm ELTRAN wafers using same process as 6 and 8 inches wafers. The key technology of 300mm ELTRAN wafers was the water jet splitting process because it was difficult to control the porous Si thickness uniformity and water jet conditions. But the porous Si layer was formed at 10um +/- 0.66um thickness, and splitting was succeeded with almost same conditions as of 8 inches process. Figure 11 shows the split 300mm wafers. The left one is the seed wafer. The black color indicates a remaining 1st porous Si layer, and the right one is the split and porous Si etched ELTRAN wafer. Its SOI thick uniformity is very good as shown in figure 12.

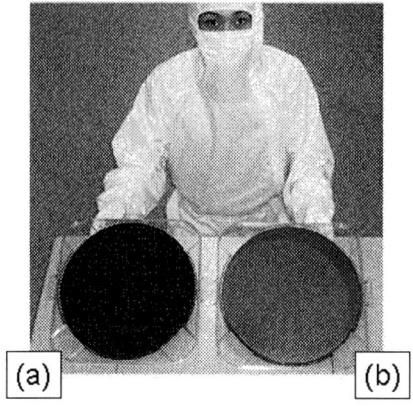

Figure 12. Split 300mm seed wafer and ELTRAN wafer

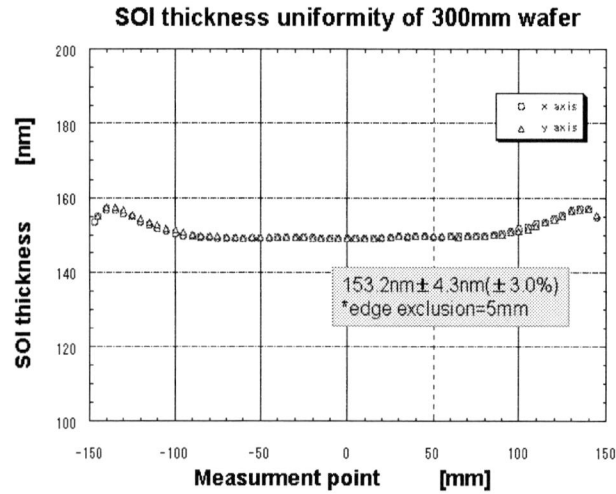

Figure 13. SOI thickness uniformity of 300mm ELTRAN

CONCLUSIONS

We designed and made the fully automated anodization system for mass production. And anodization conditions were optimized by simulation using finite element method. A water jet splitting machine and the other unique machines have been also developed, and by using these machines, we are producing 6 and 8 inches ELTRAN wafers in mass production bases. 300mm ELTRAN wafers were also produce successfully by same process as 8 inches ELTRAN wafers.

ACKNOWLEDGEMENTS

The authors would like to thank Mr. Sasaki and Mr. Oguri for useful simulation and discussions. And we also thank Mr. Matsumura, Mr. Watanabe and Mr. Nozawa for important experiments and analysis of anodization.

REFERENCES

(1) K.Imai, Solid-State Electron, vol.24, p.159, 1981

(2) L.T.Canham, Appl. Phys. Lett.**57**(10), 1990, p1046

(3) T.Yonehara, K.Sakaguchi and N.Sato, Appl. Phys. Lett.**64**(16),1994, p2108

(4) K.Sakaguchi, N.Sato, K.Yamagata, Y.Fujiyama and T.Yonehara, Extended Abstract of the 1994 Int. conf. on SSDM, 1994, p259

(5) K.Sakaguchi, K.Yanagita, H.Kurisu, H.Suzuki, K.Ohmi and T.Yonehara, Proc. 195th Int. SOI Symposium, vol.99-3, p.117-121, Electrochemical Society, (1999)

Mat. Res. Soc. Symp. Proc. Vol. 681E © 2001 Materials Research Society

Atomic-Layer Cleaving and Non-contact Thinning and Thickening for Fabrication of Laminated Electronic and Photonic Materials

Michael I. Current, Shari N. Farrens, Martin Fuerfanger, Sien Kang, Harry R. Kirk,
Igor J. Malik, Lucia Feng, Francois J. Henley, Silicon Genesis, Campbell, CA 95008, USA

Abstract

An innovative suite of layer transfer technologies, collectively called the NanoCleave™ Process, includes a non-porous cleave plane utilizing a compressive strain layer, growth of a high purity, crystalline device layer, plasma activation coupled with vacuum bonding, room-temperature cleaving along an atomically flat plane and a variety of post-cleave CVD processes to thicken or thin the device layer to a desired final thickness is described. Applications of this process include fabrication of SOI wafers containing Si and SiGe alloy device layers.

1. Introduction

Use of innovative approaches for enhanced bond strengths and engineered cleave plane structures result in processes for layer transfer of electronic and photonic materials for efficient fabrication of laminated electronic and photonic structures. These structures include Silicon-on-Insulator (SOI) wafers for electronic devices and a variety of optical signal couplers, routers and sensors. The ability to form cleaved SOI layers containing single or multi-layers Si and SiGe alloys with Angstrom-level surface finish after room-temperature separation provides an opportunity to bond new crystalline structures directly onto transferred layers without the need for mechanical polishing, etching and/or high-temperature annealing of damaged layers.

Applications for multi-level structures include integral fabrication of SOI wafers with multi-layer high-mobility channel structures, multi-level SOI wafers for fabrication of self-aligned dual-gate CMOS and a variety of optical structures. Layer thickness for applications of SOI materials range from below 10 nm to several microns for the device layer (Si or SiGe alloy) and from 50 nm to several microns for the buried dielectric (usually SiO_2 or nitrided oxides, called "buried oxide" or "BOX" layers) as illustrated in Fig. 1.

Methods to fabricate these multi-layer structures have evolved from epitaxial growth of Si on sapphire and grind-back of bonded layers (in the 1970's) to direct implantation of oxygen (SIMOX) and etching of bonded wafers (from the 1980's) to bonded processes that exploit fracture of porous layers by thermal (Ion/Smart-cut) or mechanical (Eltran) stress (in the 1990's). The recent development of controlled cleaving techniques using engineered SiGe strain layers (NanoCleave™) in conjunction with advances in layer bonding and thickness modification results in marked improvements in film thickness control, film and surface quality and production efficiency over earlier methods.

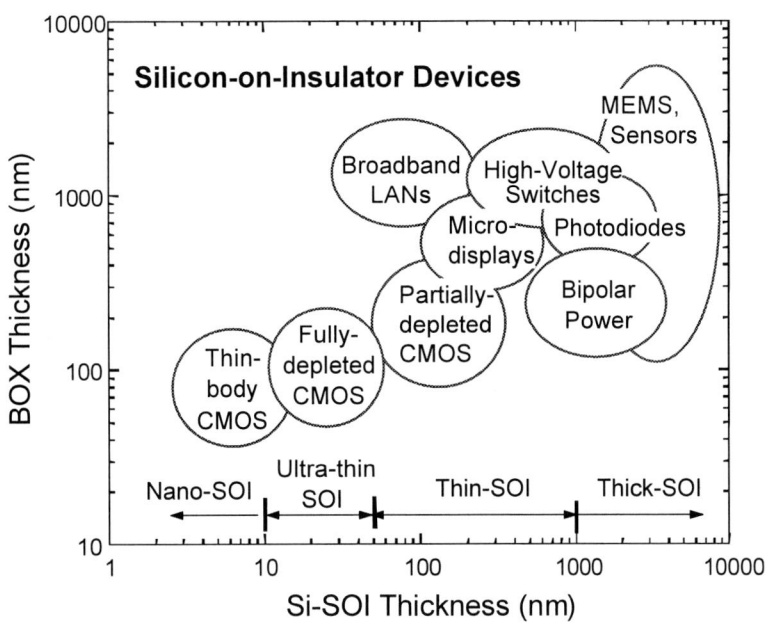

Figure 1. Layer Thickness and applications for SOI materials.

2. NanoCleave™ Layer Transfer

The NanoCleave™ layer transfer process, so named because it results in layer separation of Si with a surface roughness of considerably less than 1 nm, is a layer transfer process which includes (1) formation of a strain-layer cleave plane and high-purity epitaxial Si device layer on a "donor" wafer, (2) thermal oxidation of a "handle" wafer to form the BOX layer, (3) bonding of the donor and handle wafers, (4) separation (Atomic-layer cleaving) of the wafers within the cleave plane leaving the Si device layer attached to the BOX/handle wafer, (5) optional post-cleave CVD processing to increase or decrease the device layer thickness and smooth the final surface to below Ångstrom level roughness (Fig. 2).

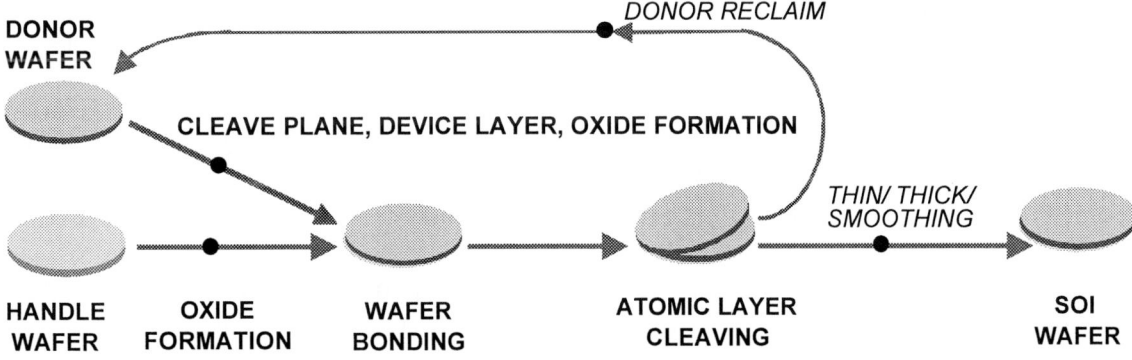

Figure 2. Schematic process flow for NanoCleave™ layer transfer formation of SOI materials.

163

2.1 Cleave Plane Engineering

Formation of a compressive strain layer by epitaxial growth of a non-dislocated SiGe alloy layer on Si followed by growth of a crystalline device layer (Si or Ge-rich multilayers) can be used to separate the upper device layer though controlled cleaving within the SiGe alloy layer. In sharp contrast to earlier reported methods of layer fracture, the cleave plane region contains no porous regions, inclusions, or voids, as shown in Fig 3.

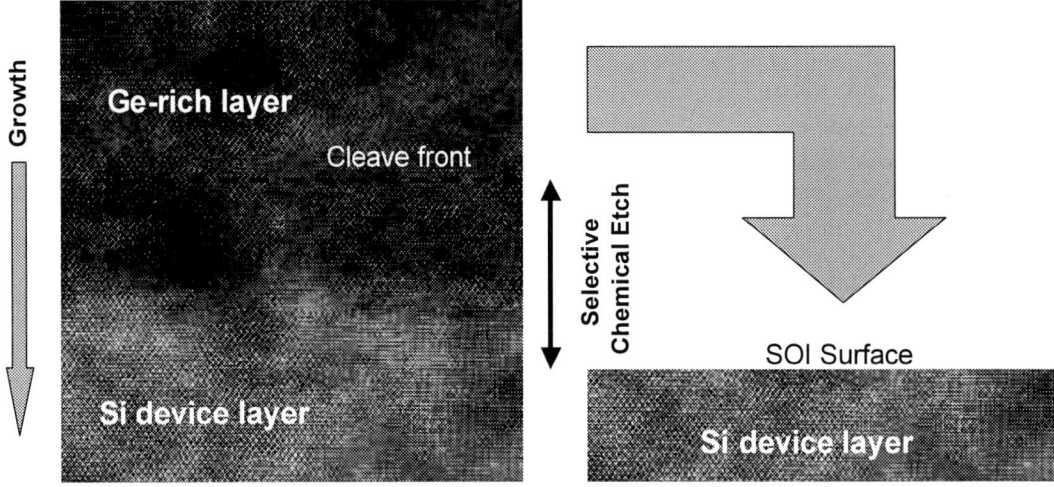

Figure 3. High-resolution TEM image of the boundary area of a Ge-rich, epitaxial strain layer and a crystalline Si device layer in a state prior to atomic-layer cleaving. The atomic-layer cleaving will occur within the Ge-rich layer. The residual Ge-rich layer on the as-cleaved wafers are subsequently removed through a selective etch step. Note that the cleave plane region is continuous on an atomic scale and contains no voids or other porous structures.

The resulting separated surface has a surface roughness, as measured by AFM profiling, of 2 Ångstroms or less. The surface finish has been found to be smooth, uniform and AFM imaging shows no notable topography anomalies over measurement scales from 2 to 30 um [1].

2.2 Plasma-activated dry bonding

Layer transfer techniques require that the cleave plane (1) have an engineered stress profile to guide a propagating fracture along the desired cleave direction (to prevent 3-dimensional fracture) as well as (2) be weaker than the bonded layer (to prevent delamination of the bonded layer during cleaving). The second criterion requires in turn that the bonding process result in high bond strengths without serious deleterious effects, such as plastic deformation of the bonded materials during high-temperature annealing. Exposure of the wafer surfaces to a low-energy plasma prior to bonding can produce as-bonded, "contact", bond strengths that are more than an order of magnitude higher than bonds obtained following soaks in wet chemistries, such as $NH_4OH:H_2O_2:H_2O$ [2]. The bond strength of "plasma activated" surfaces can be further increased by another order of magnitude through modest (200 to 300 C) thermal annealing, resulting in bond strengths that are achieved following wet-chemistry bonding only after annealing above 800 C.

164

The van der Waals contribution to the contact bond strength can be modeled by estimating the dielectric polarization of the contacting layers [3]. Model layers for the case of wet-chemistry and plasma activated, "dry" bonding of oxide coated Si wafers are shown in Fig. 4. The key variables in this model are the thickness of the interfacial moisture and other dielectric layers (including the plasma activated surfaces) and the dielectric constants of the layers.

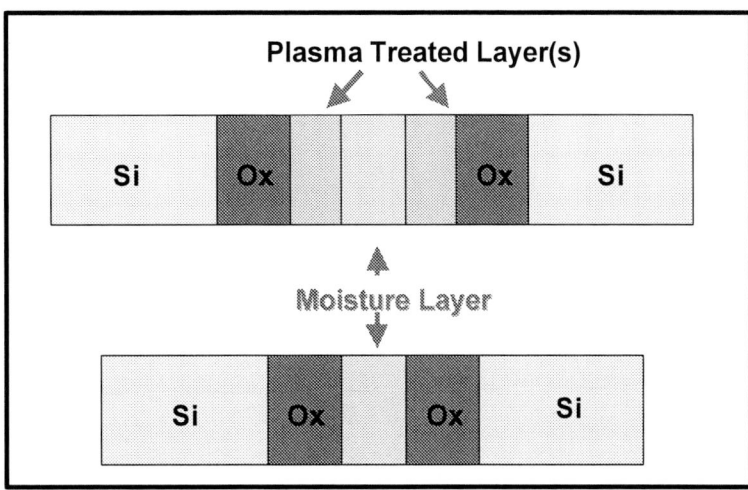

Figure 4. Model layers for van der Waals attraction model for plasma activated, dry bonding (above) and wet-chemistry bonding (below) of oxide covered Si wafers.

The net disbonding pressure [2,3] is calculated as a sum of reflection and transmission paths and frequencies of the electric potentials in the bonded layers. The resulting expression for disbonding pressure, Π, is in the form:

$$\Pi \text{ (MPa)} = (h/16\pi^3)*(\Sigma_{\omega,\text{layers}}[f(\varepsilon_i)*(1/(\Sigma a_i)^3)] \sim f'(\varepsilon'_i)*(1/(a'_i)^3) \quad (1)$$

where h is Planck's constant, $\Sigma_{\omega,\text{layers}}$ is a sum over all frequencies and closed loop paths of transmitted and reflected electric potentials in the bonded multi-layers for a product of $f(\varepsilon_i)$, which are products and sums of layer dielectric constants, and $(1/(\Sigma a_i)^3$, which is the inverse cube of sums of the individual layer thickness, a_i. The leading term of the sum, $f'(\varepsilon'_i)*(1/(a'_i)^3)$, is dominated by the effect of the thinnest and most polarized layer. In the case of the wet-chemistry bond, the key layer is the interfacial moisture layer. In the case of the plasma activated bonding, the key layers are the polarized surfaces of the oxides exposed to the plasma. The much thicker interfacial moisture layers for wet-chemistry bonding results in significantly weaker contact bonds than the case of plasma-activated, "dry" bonding because of the strong effect of the inverse cubic dependence of the bond strength on the critical layer thickness. The agreement of this model with measured contact bond strengths for interfacial moisture layer thickness of 15 to 30 Å is shown in Fig. 5.

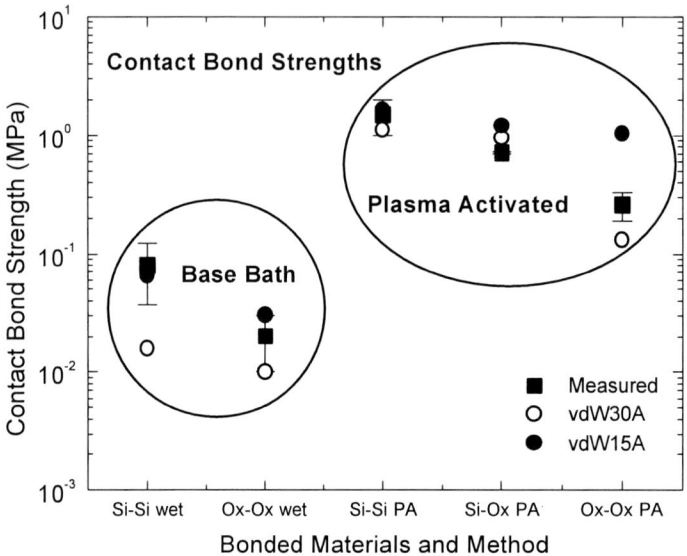

Figure 5. Contact bond strengths for Si and SiO$_2$ surfaces following base-bath soaking or oxygen plasma activation. Also shown are debonding pressures estimated for these surfaces based on the van der Waals attraction potentials and interfacial moisture layers of 15 and 30 Å.

The close agreement between the measured contact strengths and the van der Waals attraction forces is an indication that factors that could compromise good bonding action, such as surface roughness, interfacial contamination and incompatible mechanical properties, are not significant for these semiconductor processing–grade surfaces.

Apparent bond energies for plasma-activated Si<100>-Si<100> and Si-oxide interfaces (as judged by IR imaging of crack lengths with a blade-insertion test) increase after thermal annealing between 200 and 300 C to >5 J/m^2 (corresponding to bond strengths of >5 MPa) [2]. These bond energies exceed the 2 to 3 J/m^2 surface energies for bulk fracture along <111> and <110> directions [4] and are an indication of the low density of defect sites for fracture failure in these bonded interfaces. The development of strong covalent bonds during these low-temperature anneals provides bond strengths that are significantly higher than the cleave plane fracture energies within the strain layer, allowing for reliable separation of the wafers and layer transfer of the device layer to the handle wafer.

2.3 Room-temperature, controlled cleaving

With properly engineered cleave planes and bonded interfaces, the separation of NanoCleave™ bonded wafer pairs can be accomplished at room-temperature with modest mechanical force. A fully automated, cassette-to-cassette cleaving tool, shown in Fig. 6, contains a robotic handler (with the pickup wand visible at the upper right) for transport of the bonded pairs into the cleaving jig and removal after cleaving of SOI product and used donor wafers. The separation is accomplished in a clamshell chamber where the groove formed by the tapered edge bevels at the center plane of the bonded wafer pair is pressurized with an inert gas (usually N$_2$) through the gas feed line visible on the front of the cleaving jig. After the separation front proceeds across the bonded pair (in the time frame of ~10 seconds), the inert gas serves to keep the wafer pair from rebonding until the clamshell is opened, carrying the used donor wafer up and away on the chamber lid and leaving the layer transferred wafer in place awaiting the return of the robotic pickup arm.

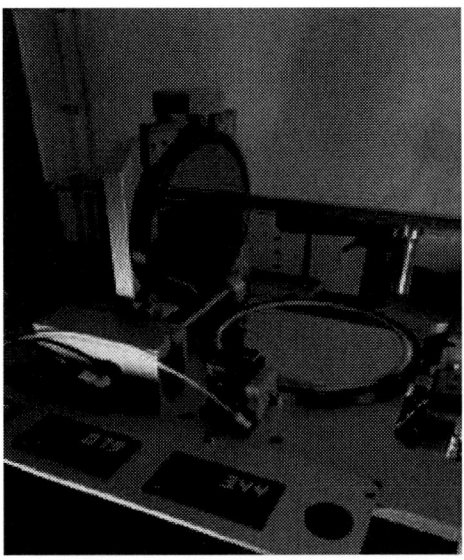

Figure 6. Fully automated, wafer separation tool used in the NanoCleave™ process. The wafer handling robot, visible in the in upper right, carries bonded wafer pairs into the separation chamber and removes layer transferred product and used donor wafers to appropriate output cassettes. The separation is accomplished at room-temperature by application of modest hydrostatic pressure by the introduction of inert gas through the gas feed line visible in the front of the separation chamber. The photo shows the bonded pair after separation with the used donor on the clamshell lid and the finished SOI wafer ready for the robotic pickup.

The progress of the cleave front can be monitored by an acoustic pickup coupled to the bonded pair. The acoustic signal may be used to confirm the completion of the layer separation operation. The frequency spectrum of the acoustic signal contains characteristic peaks which shift depending on the molecular weight of the gas used in the separation process (Fig. 7). It is anticipated that analysis of these acoustic signals may be exploited in process control procedures for automated, high-volume manufacturing.

2.4 Post-cleave smoothing and thickness tuning
As the applications for SOI-type materials proliferate (see Fig. 1) and prior to the development of industry-wide standard wafer configurations, there is a need to fabricate small lots of SOI wafers with a wide variety of Si-SOI thickness while still maintaining precise control on the individual wafer thickness and operating with good manufacturing efficiency. Post-cleave processing in commercial epi-deposition chambers has demonstrated a range of final Si-SOI thickness from 10 nm [5] to several microns using as-cleaved SOI in the range of 100 to 200 nm Si-SOI thickness.

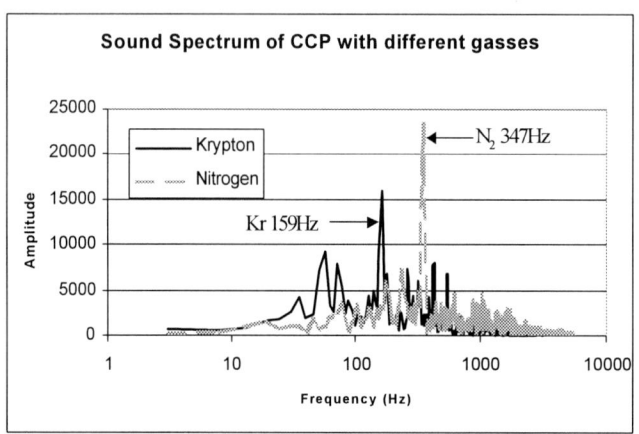

Figure 7. Acoustic spectra during room-temperature separation using Kr and N_2 gas flows.

Post-cleave process for layer thickness adjustment include: (1) Si layer thinning using etching processes [5] and layer thickening by growth of additional Si or SiGe multi-layers. The layer thickness control for thickened wafers is generally better than 0.5% with a within-wafer thickness uniformity of 2% or better (Fig. 8). An additional benefit of these post-cleave layer thickness adjustments is that the surface finish is improved to the range of 0.6 to 0.9 Å (as measured by 2x2 um AFM tapping mode scans).

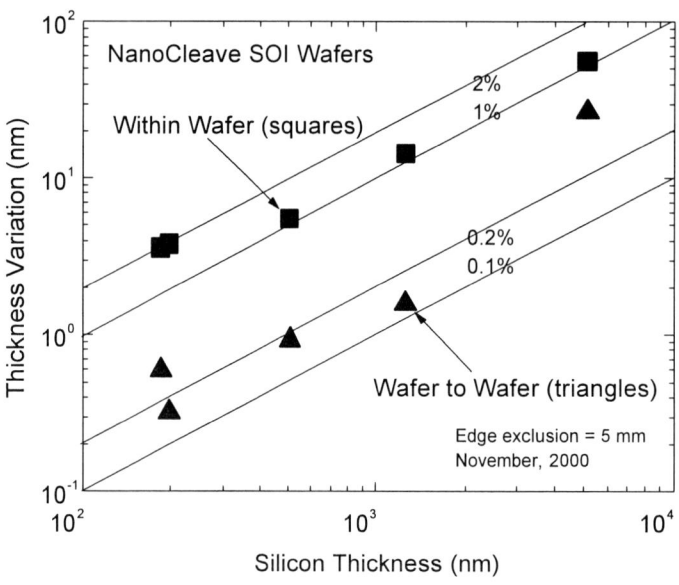

Figure 8. Si-SOI layer thickness uniformity and average layer thickness control for post-cleave thickening of as-cleaved SOI wafers with cleaved 200 nm Si-SOI layers.

4. New Applications: SiGe multilayers

Ge-rich Si layers in SOI-type materials are of increasing interest for the development of high-mobility CMOS transistors and to extend the range of efficient absorbtion of photonic signals with Si-like photodiodes from 0.8 um for Si into the near infra-red range of 1.3 to 1.55 um used for long-range fiber-optic cable links. The NanoCleave SOI process is well-suited to modifications which include Ge-rich alloys into the device layer. An example where Ge-rich layers were incorporated into the donor wafer and transferred to an oxidized handle wafer is shown in Fig. 9. The Ge- layers were deposited with a triangle shape with a peak at 25% Ge and a step structure with 27, 17 and 9% Ge ledges. The Ge layers were sandwiched between a 40 nm Si layer and a 2 nm top Si cap. The Ge-rich multi-layer was bonded and cleaved and annealed in a multi-ramp cycle at 800 to 900 C for ~1 hr to simulate a source/drain dopant anneal. Substantial diffusion from the Ge layers was observed after the dopant anneal cycle, with the final peak concentrations dropping from 25-29% to 16-19% Ge. High-resolution TEM images showed no indication of extensive dislocation formation in these annealed Ge-rich multi-layers.

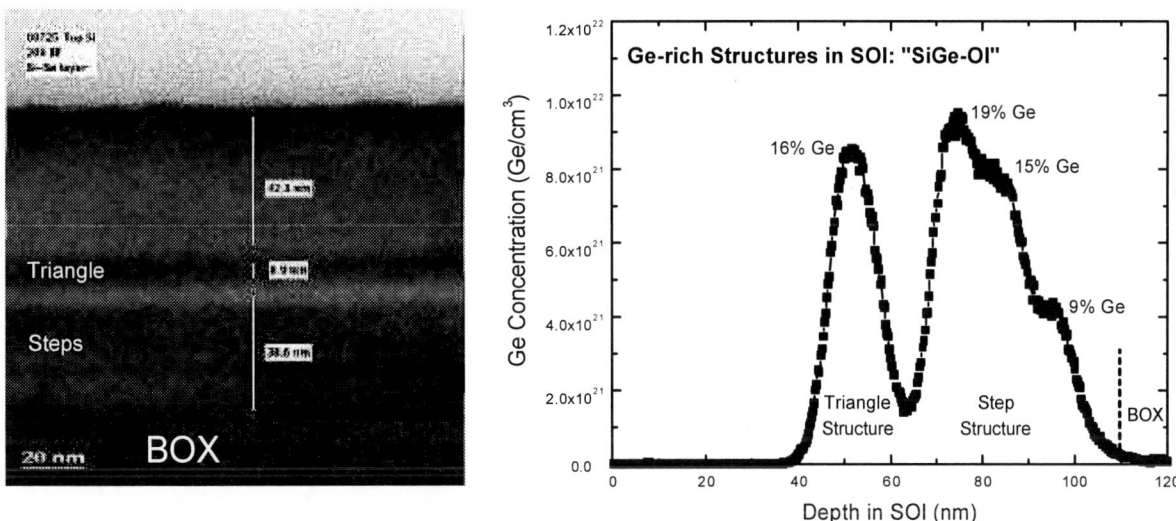

Figure 9. TEM and SIMS profiles of a 100 nm thick SiGe-SOI layer formed by transferring Ge-rich multi-layers onto an oxidized handle wafer. The SiGe-SOI was annealed at 800-900 C for ~1hr to simulate a particular CMOS dopant activation thermal cycle.

Ge-rich multi-layers can also be added to post-cleave Si-SOI wafers. An example of a three-layer Ge-rich structure grown on a 200 nm Si-SOI layer is shown in Fig. 10. These thin Ge layers, with peak Ge content in the range of 5 to 20 %, were grown between 5 nm Si spacers.

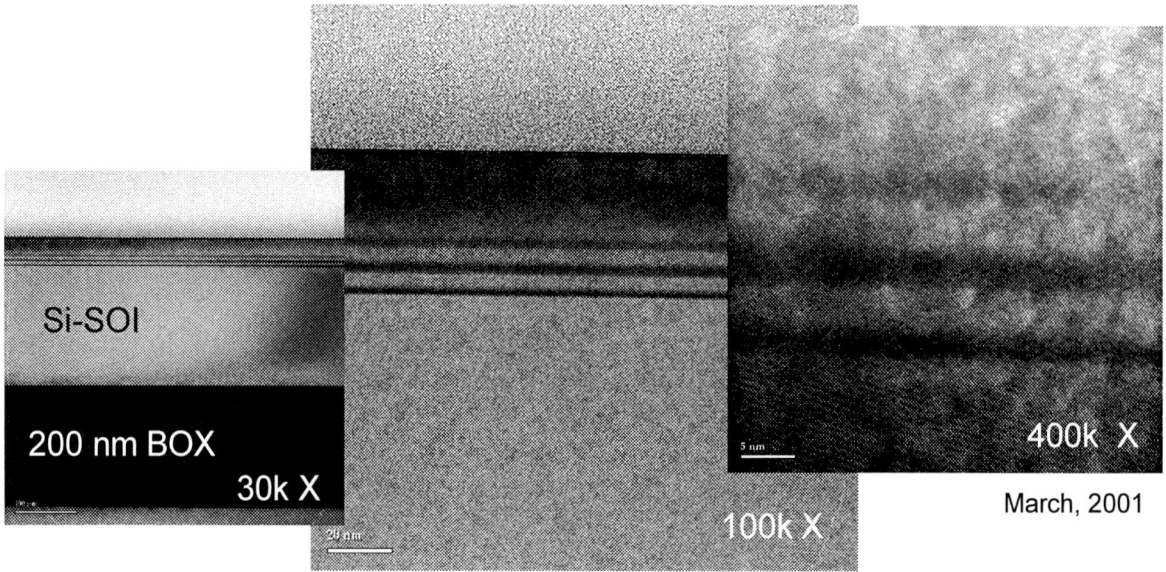

Figure 10. TEM images of three Ge-rich layers separated by 5 nm Si layers grown on a 200 nm Si-SOI layer with a 25 nm Si cap and a 200 nm buried oxide layer.

5. Summary

The development of efficient and flexible technologies [1,6,7] for layer transfer fabrication of SOI and other forms of laminated materials using a suite of techniques, featuring atomic layer cleaving within a non-porous strain layer, plasma-activated bonding and post-cleave layer treatments, opens new possibilities for multi-layer materials including various paths towards 3-D integration of electronic and photonic systems.

Acknowledgements

The authors acknowledge the contributions to the development of the NanoCleave™ process from the Silicon Genesis fab group, Nathan Cheung at UC Berkeley, Steve Bedell, now at IBM, and Bill En, now at AMD.

References:

1. M.I. Current, I.J. Malik, M. Fuerfanger, M. Korolik, S. Kang, H. Kirk, M. Fang, S.N. Farrens, F.J. Henley, in "Silicon-on-Insulator Technology and Devices X", eds. S. Cristoloveanu et al., Electrochemical Society, Proc. 2001-3 (2001) 75-78.
2. S. N. Farrens, J.R. Dekker, J. K. Smith, B. E. Roberds, J. Electrochem. Soc. **142** (11) (1995) 3949-3955.
3. D. Langbein, J. Adhesion **3** (1972) 213-235.
4. S.G. Roberts, in "Properties of Crystalline Silicon", ed. R. Hull, INSPEC (1999) 144-148.
5. A-L. Thilderkvist, S. Kang, M. Fuerfanger, I.J. Malik, 2000 IEEE Inter. SOI Conf., October, 2000, 12-13.
6. M.I. Current, I.J. Malik, S.W. Bedell, H. Kirk, M. Korolik, S. Kang, F.J. Henley, in "High Purity Silicon VI", eds. C.L. Claeys et al., Electrochemical Society, Proc. 2000-17 (2000) 516-523.
7. M.I. Current, S.W. Bedell, I. J. Malik, L.M. Feng, F.J. Henley, Solid State Technology, (July, 2000), 66-77.

Mat. Res. Soc. Symp. Proc. Vol. 681E © 2001 Materials Research Society

Orientation and Boron Concentration Dependence of Si Layer Transfer by Mechanical Exfoliation

Kimmo Henttinen, Tommi Suni, Arto Nurmela, Veli-Matti Airaksinen[1], Ilkka Suni and S.S Lau[2]

VTT Electronics, FIN-02044, Finland.
[1]Okmetic Oyj, FIN-01510 Vantaa, Finland.
[2]Department of Electrical and Computer Engineering, University of California, San Diego, La Jolla, CA 92093-0407, U.S.A.

ABSTRACT

Mechanical exfoliation strength has been measured in hydrogen implanted <100>, <111> and <110> oriented Si wafers using the crack opening method. The bonding temperature required for exfoliation increases in the order <100>, <111> and <110>. The same method has been applied to study the influence of boron doping on mechanical exfoliation in <100> Si wafers. The required bonding temperature to exfoliate mechanically decreases with increasing doping level independent of the electrical activation of boron. The enhanced crystallization rate of boron doped Si is suggested as a plausible explanation for the result.

INTRODUCTION

Low-temperature bonding and layer transfer processes are promising techniques for three dimensional integration of electronic, optical and micromechanical devices. Hydrogen implantation induced silicon layer transfer is a well-documented method to form silicon on insulator (SOI) and other device structures [1]. When the H-implanted and bonded wafer is annealed at moderately high temperatures of 400-600°C, the exfoliation of the implanted wafer takes place due to hydrogen pressure built-up in the microcavities located near the projected range of hydrogen. It has been shown that the H-implanted silicon can also be exfoliated by mechanical means after low-temperature bonding and subsequent bond annealing at 150-200 °C [2,3]. As an implanted and bonded wafer pair is subjected to mechanical splitting forces, fracture takes place along the path of least resistance. If the implanted region is rendered weaker than the bonded interface, the fracture propagates within or near the implanted zone, instead of at the bonded interface. Mechanical exfoliation has the advantage of producing a smooth split surface and an intrinsically low-temperature process for the matching of dissimilar materials. Recently, the mechanical splitting process has been reported for patterned implanted Si wafers [4].

Many physical and chemical phenomena in silicon are dependent on crystal orientation and doping concentration. The objective of this work was to study the effect of silicon crystal orientation and boron concentration on the mechanical exfoliation process. Similar studies have been carried out to investigate blister formation associated with the thermal splitting process [5,6,7]. We have carried out direct measurements on hydrogen induced weakening of the implanted layer using the crack opening method [8]. The implanted and split layers were characterized by Rutherford backscattering spectrometry (RBS) and ion channeling, atomic force microscopy (AFM) and optical methods.

EXPERIMENTAL

In this work <100>, <111> and <110> oriented p-type 100 mm Si wafers with a resistivity of 1-35 Ωcm were used as donors for layer exfoliation. A 30 nm thick screen oxide was grown on these wafers before ion implantation. The handle <100> oriented Si wafers were thermally oxidized to a thickness of 500 nm of SiO_2 to make a silicon dioxide to hydrophilic silicon bond (oxide/Si bond). The donor wafers were implanted with H_2^+ at 100 keV at a dose of 4.5×10^{16} or 5.0×10^{16} H_2/cm^2. Prior to hydrogen implantation some of the <100> oriented donor wafers were implanted with B^+ at 175 keV to a dose ranging from 10^{13} cm^{-2} to 3×10^{15} B/cm^2. No electrical activation was carried out for implanted boron. During the hydrogen implantation, the beam current was kept below 100 μA to avoid uncontrolled heating of the wafer. To minimize the channeling effects the wafer surface was tilted 7° with respect to the incident beam during the implantation. To study the role of the electrical activity of the boron, a control sample was prepared by implanting hydrogen to a donor wafer with a boron doped epitaxial layer. The boron doping concentration of the epilayer was 7.83×10^{19} cm^{-3}, which is approximately one half of the peak concentration of the boron obtained at the implantation dose of 3×10^{15} B/cm^2 (see figure 1). After the implantation the wafers were cleaned with Piranha and RCA solutions and the screen oxide was removed. The implanted donor wafers were subsequently bonded to oxidized handle wafers. A low-temperature, plasma enhanced Si bonding process was used to obtain strong bonding already at 200°C [3]. The bonded wafer pairs were annealed in a box furnace under N_2 in two steps: first for 2 hours at 100°C and then 2 hours at 180-400°C. After annealing, the wafers were cut with a dicing saw to rectangular slices. The strength of the implanted layer was measured by the crack opening method in air [8]. A blade was inserted between the wafers and a crack propagating along the implanted region was formed when

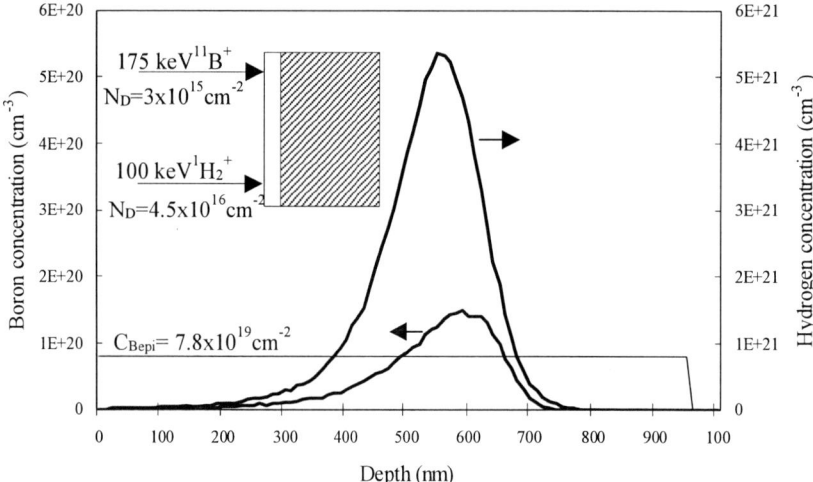

Figure 1. Implanted boron and hydrogen profiles simulated with SRIM 2000 [9]. Only the largest boron dose is shown. The constant boron concentration of the epilayer is given as reference.

the bonded interface was stronger than the H-implanted layer. The blade was inserted along the [110] direction for <100> and <111> Si wafers and along [111] for <110> Si wafers. The surface energy of the implanted layer was calculated based on the crack length measurement [8]. The split surfaces were studied with a Digital Instruments D3100 atomic force microscope (AFM) using silicon tips in the tapping mode. The thickness of the exfoliated layer was measured by optical reflectometry (Nanospec/AFT 4150) and Rutherford backscattering spectrometry (RBS). To improve the depth resolution of the RBS measurements, we used a 2 MeV ^{4}He$^+$ beam at 40° off the normal incidence. The implantation damage was characterized by 2 MeV ^{4}He$^+$ backscattering in the channeling mode. The channeling measurements were done with unbonded donor wafers annealed at 100°C for 2 hours after the implantation.

RESULTS

The strength of the hydrogen-implanted layer measured with the crack opening method after isochronal (2 hrs) annealing steps in the temperature range 200-400 °C is depicted in figure 2 for <100>, <111> and <110> oriented silicon. The required annealing temperature for layer splitting is lowest for <100> silicon and highest for <110> silicon. The AFM-measurements of the split surface yielded an average surface roughness (R_a) of 3-4 nm for all three orientations.

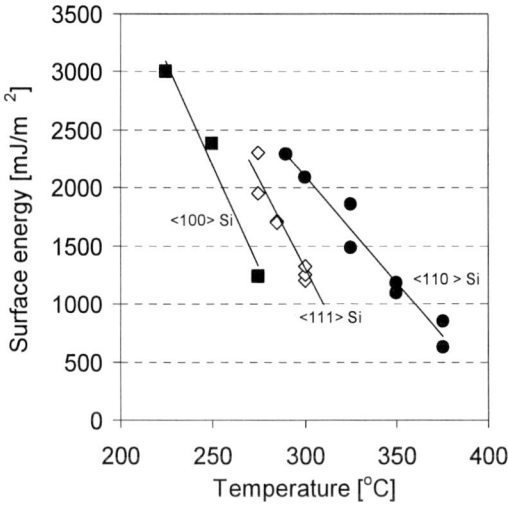

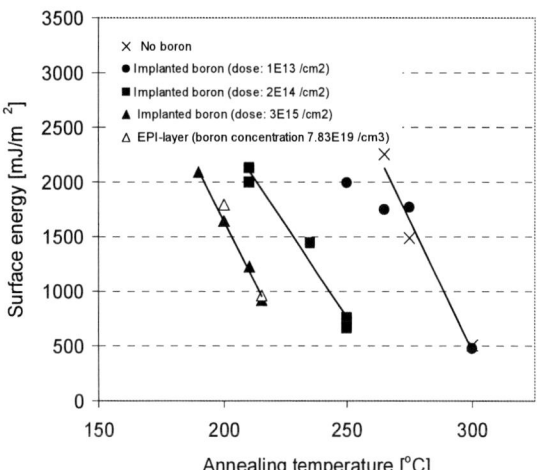

Figure 2. The strength of the hydrogen implanted layer in <100>, <111> and <110> oriented Si wafers. The surface energy was measured using the crack opening method in air. The samples were implanted with 5×10^{16} H$_2$/cm^2 at 100 keV. The annealing time was 2 hours.

Figure 3. The measured surface energy for different boron doping schemes in H-implanted and bonded Si. The boron implantation energy was 175 keV. The boron doses are 1×10^{13} cm^{-2} (circles), 2×10^{14} cm^{-2} (squares) and 3×10^{15} cm^{-2} (closed triangles). Results for undoped samples (crosses) and doped epi-layers (open triangles) are also shown. The hydrogen implantation was done with H$_2^+$ at 100 keV with a dose of 4.5×10^{16} cm^{-2}. The annealing time was 2 hours.

The thickness of the exfoliated layer was 478 nm for all samples suggesting that the fracture propagates at the same depth of the implanted layer if the implantation conditions are kept constant and the channeling effects are avoided during the implantation.

The influence of the boron implantation on the mechanical exfoliation is shown in figure 3. The strength of the H-implanted layer was found to decrease with boron doses $> 10^{13}$ cm^{-2}. Consequently, the temperature required for splitting decreases with increasing boron dose. For the studied dose range from 1×10^{13} B/cm^2 to 3×10^{15} B/cm^2 the temperature difference is almost 100 °C. Figure 3 also shows that the heavily boron doped epi-layer and the sample implanted with the highest boron dose follow similar exfoliation behavior. This suggests that the strength of the H-implanted layer is unaffected by the electrical activity of the boron prior to hydrogen implantation.

The exfoliation depths in boron doped samples were determined from RBS measurements carried out in the random mode. The samples implanted with 1×10^{13} and 2×10^{14} B/cm^2 exfoliate at the depth of approximately 478 nm, identical with the exfoliation depth of the undoped sample. The thickness of the exfoliated layer in both the heavily boron doped epilayer and the sample implanted with 3×10^{15} B/cm^2 was 459 nm. To investigate the differences in the damage distributions caused by the ion implantation, RBS measurements were also carried out in channeling mode. The results are presented in figure 4. A large damage peak is observed at a slightly shallower depth than the projected range of hydrogen. The defect density appears to be largest for the undoped sample and smallest for the highly boron doped epilayer. The heavily boron implanted sample has a wider defect peak presumably because of the defects induced by the boron implantation.

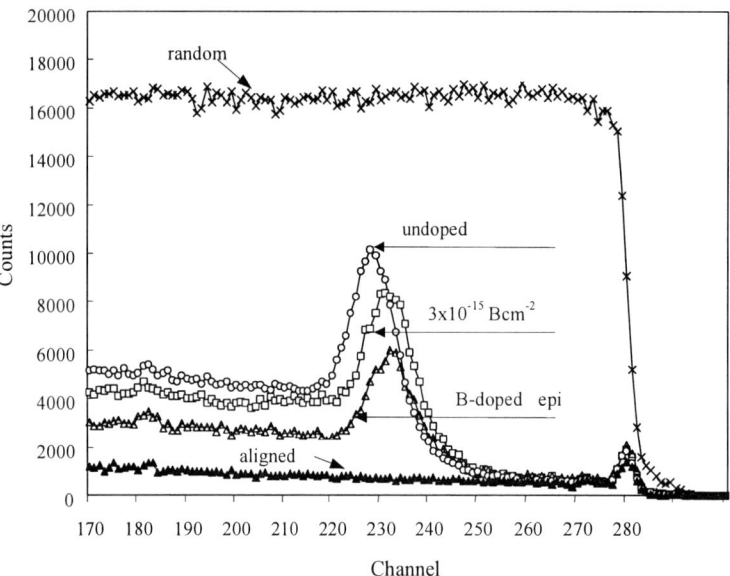

Figure 4. 2 MeV ^{4}He$^+$ channeling spectra of the hydrogen implanted layer (4.5×10^{16} H$_2$/cm^2) for the various boron doping schemes. After the implantation the samples were annealed at 100°C for 2 hours. The random and aligned spectra from a virgin sample are shown as reference.

174

DISCUSSION

High dose hydrogen implantation is known to create vacancies, interstitials and hydrogen complexes by breaking Si-Si bonds [10]. In silicon, the strain increases linearly with the implantation dose until the amorphization threshold is reached or the strain is relieved by extended defects. A specific class of hydrogen-stabilized plate-like defects arise from the coordinated formation of Si-H bonds. These platelets are believed to act as a nucleation sites for hydrogen filled cavities and microcracks [10]. They decrease the fracture strength of the implanted layer making the mechanical exfoliation possible. The strength of the implanted layer depends on the density and size of the microcracks and platelets. It's known that during annealing the hydrogen filled cavities grow in size and reduce in density while their overall volume remains constant [11]. In mechanical splitting process, the stress field of the crack favors the propagation of the fracture along the plane parallel to the wafer surface. The AFM-measurements show that the fracture appears to propagate randomly through many planes following the path of minimum fracture energy inside the implanted layer. The exfoliation temperatures for <100>, <111> and <110> follow the same order as recently reported for surface blistering in Si [6]. The hydrogen microcracks form more sluggishly in <110> and <111> Si than in <100> bonded Si.

The presence of the boron reduces significantly the strength of the H-implanted layer enabling mechanical exfoliation at temperatures below <200°C. This observation is comparable to the surface blistering results obtained by Tong et. al.[5] The activation of the boron before the H-implantation appears to have no influence on the temperature required for mechanical layer transfer. Comparable results were obtained with implanted but not activated and epitaxially grown layers (see figure 3). It has been shown that the stress introduced by implanted or substitional boron is not involved in lowering the splitting temperature [5]. We also believe that the damage due to boron implantation is not the primary reason for the reduced exfoliation temperature, because B+H co-implanted samples have less damage than the samples implanted with hydrogen only. The H-implanted epilayer shows similar reduction in hydrogen induced damage and exfoliation temperature. The disorder induced by the implanted hydrogen was found to decrease with increasing boron concentration. The result is consistent with the well-established fact that both self-diffusion and recrystallization rates in Si are accelerated by a high concentration of boron. As a result, fast recovery takes place at the trailing edge of the damage distribution and the peak appears to move closer to the surface when the boron concentration is > 10^{19} cm^{-3}. The depth of exfoliation is also smaller in the samples with high boron concentration. Our experimental results suggest that the maximum compressive stress at this hydrogen dose is within the recovered zone in the case of the boron doped samples. Our findings agree with previously reported results where above a critical damage the fracture toughness of the damaged layer increases [12].

CONCLUSIONS

Our experiments show that the crack opening method can be used for measuring the exfoliation strength of the hydrogen-implanted and bonded silicon. The crack opening method directly probes the evolution of hydrogen-filled microcavities leading to the fracture. It is shown that the exfoliation temperature depends on the Si crystal orientation. It is lowest for <100> and highest for <110>oriented Si, in agreement with recent results on hydrogen induced blistering in

Si. Implantation and doping with boron reduce the strength of the hydrogen-implanted region. This is consistent with reported results on surface blistering in B+H co-implanted Si. We suggest that this is largely due to the enhanced recrystallization rate in boron doped silicon leading to a lower fracture strength in the regrown zone.

ACKNOWLEDGMENTS

The authors would like to thank M. Nastasi and T. Höchbauer from Los Alamos National Laboratory (New Mexico, USA) for helpful comments. UCSD acknowledges NSF for support.

REFERENCES

1. M. Bruel, Electron. Lett. **31**, 1201 (1995).
2. W.G. En, I. J. Malik, M. A. Bryan, S. Farrens, F. J., Henley, N. W. Cheung, C. Chan, IEEE International SOI Conference Proceedings, Stuart, FL, USA, 5-8 Oct. 1998. New York, NY, USA: IEEE, 1998. p. 163-164.
3. K. Henttinen, I. Suni and S. S. Lau, Appl. Phys. Lett. **76**, 2370 (2000).
4. C. H. Yun, N. W. Cheung, Proc. 13th Conf. Ion Implantation Technology, Alpbach, Austria, 17-22 Sept. 2000.
5. Q.-Y. Tong, R. Scholz, U. Gösele, T.-H. Lee, L.-J. Huang, Y.-L. Chao and T. Y. Tan, Appl. Phys. Lett. **72** , 49 (1998).
6. Y. Zheng, S. S. Lau, T. Höchbauer, A. Misra, R. Verda, X.-M. He and M. Nastasi, J. Appl. Phys. **89**, 2972 (2001).
7. T. Höchbauer, K. C. Walter, R. B. Schwarz, M. Nastasi, R. W. Bower and W. Ensinger, J. Appl. Phys. **86**, 4176 (1999).
8. Q.-Y. Tong, U. Gösele, Semiconductor Wafer Bonding, Electrochemical Society Series, Wiley, 1999.
9. J. F. Ziegler, J. B. Biersack and U. Littmark, The Stopping and Range of Ions in Solids (Pergamon, New York, 1985).
10. M. K. Weldon, V. E. Marsico, Y. J. Chabal, A. Agarwal, D.J. Eaglesham, J. Sapjeta, W. L. Brown, D. C. Jacobson, Y. Caudano, S. B. Christman and E. E. Chaban, J. Vac. Sci. Technol. B **15** (4), 1065 (1997).
11. J. Grisolia, G. Ben Assayag, A. Claverie, B. Aspar and C. Lagahe, Appl. Phys. Lett. **76**, 852 (2000).
12. M. Nastasi, T. Höchbauer, A. Misra, R.D. Verda, J.W. Mayer, S.S. Lau, K. Henttinen, I. Suni, presented at the 16th International Conference on the Application of Accelerators in Research and Industry, CAARI 2000, Nov. 1-4, 2000, Denton, Texas, USA.

Mat. Res. Soc. Symp. Proc. Vol. 681E © 2001 Materials Research Society

Structured monocrystalline Si thin-film modules from layer-transfer using the porous Si (PSI) process

Auer Richard and Brendel Rolf
Bavarian Center for Applied Energy Research (ZAE Bayern)
Am Weichselgarten 7, D-91058 Erlangen, Germany

ABSTRACT

We demonstrate a novel technique for fabricating monolithically series connected solar modules from surface structured thin monocrystalline Si films that we prepare by layer transfer using porous Si (PSI process). The novel series connection technique bases on reactive ion etching of the silicon film in a microwave plasma prior and after layer transfer. The module has an area of 25 cm^2 and consists of 5 unit cells that have a film thickness of 16 μm. We measure an open-circuit voltage of 3028 mV and a confirmed efficiency of 9.9%. The Si film has a randomly textured surface for light trapping.

INTRODUCTION

Cost reduction and higher conversion efficiencies are the major objectives of the photo-voltaic-R&D work. Monocrystalline silicon wafers of typically 300 μm in thickness achieve high conversion efficiencies, but sawn wafers from ultra-pure single crystals are an energy-intensive and costly material. For cost reduction, the fabrication of thin crystalline Si cells with thickness values of 3 to 25 μm is under heavy investigation. Thin-films in this thickness range require a supporting carrier. Unfortunately epitaxial growth on low-cost carriers such as glass is a very difficult task, that generally leads to crystal defects which reduce the conversion efficiency of the device. Layer transfer using porous Si is a technique to fabricate high quality thin-film cells on glass. The ELTRAN process applies epitaxy on porous Si and transfers a planar film to a carrier by bonding [1]. A planar thin Si film does however not absorb the solar light sufficiently. We therefore introduced the porous Si (PSI) process [2], that fabricates monocrystalline thin-film cells with a surface structure by performing epitaxy on structured porous Si.

Using layer transfer with porous Si, Tayanaka et al. achieved 12.5% solar cell efficiency [3] with a 12-μm-thick and planar Si film. Rinke et al. reported 14.0% efficiency [4] for a 24.5-μm-thick and planar Si film. Both authors apply sophisticated high-temperature processing steps including photolithography. At ZAE Bayer we develop low-cost processes for thin-film solar cells. Using a simple processing sequence without photolithography, we reported an efficiency of 12.2% for a cell with a thickness of 16 μm that has random pyramids for light trapping [5]. Having introduced light trapping and having simplified the cell process the open question is how to fabricate a *solar module* using layer transfer techniques. Integrated series connection is required for almost all applications to enhance the output voltage and to reduce the ohmic losses of large area devices.

There are several approaches to realize an integrated series connection for thick standard Si wafers. Keller reports an open-circuit voltage of 3.43 V and an efficiency of 10.9% for a

21.0 cm^2-large monolithic module with 6 unit cells on a standard wafer (thickness > 300 μm) [6]. This techniques is not applicable to thin-film cells.

Takato et al. obtained 2.86 V from 5 unit cells fabricated from a 3-μm-thick silicon-on-insulator (SOI) layer and 10.65 V from 20 unit cells, respectively [7]. The conversion efficiency of 5.4% (for 5 unit cells) and 4.7% (for 20 unit cells) and the total size of the module is only 0.358 cm^2.

Kerst et al. reported a module with 20 unit cells [8] with 7.5 V (375 mV per cell) and 17 mA/cm^2 on a 15 μm thick and 4 cm^2-large SOI film.

Matsushita et al. report an integrated single crystalline module from 10 μm-thin Si film transferred to a plastic carrier [9]. The process apparently requires porous Si formation and oxidation for the electrical insulation of adjacent cells as well as laser ablation to open contact windows. The open circuit voltage per cell was about 550 mV. An efficiency was not reported.

Going through this list of previous work we find that non of the disclosed approaches yields thin-film modules with light trapping in a simple manner.

In this work we present the first experimental data on a novel process that we believe is simple, is applicable to thin film-cells from layer transfer, and permits efficient light trapping.

EXPERIMENTAL

Fabrication of a textured epitaxial thin-film by the porous Si process

The thin-film fabrication starts with a p$^+$-type, boron doped ($N_A > 6\times10^{18}$cm^{-3}), monocrystalline CZ-silicon substrate wafer with 4" diameter and a thickness of 525 μm. A chemical etching process generates random upright pyramids on the (100)-oriented Si surface with an average height of 3 - 5 μm. We electrochemically etch a double-layer system of porous Si in diluted HF. The first porous layer (*start layer*) has a thickness of less than 1 μm, a porosity of 20% and is etched at a current density of 5 mAcm^{-2}. The second layer (*separation or perforation layer*) is less then 300 nm thick and has a porosity exceeding 50%. The separation layer is formed at a current density of 250 mAcm^{-2}. Infrared camera imaging of the porous system reveals laterally homogeneous porous layers over the whole wafer. An annealing prior to the Si deposition for 30 min at 1100°C in the CVD reactor under hydrogen atmosphere modifies the microscopic structure of the porous layer system: In particular the surface closes [10] and the mechanical strength of the separation layer decreases [3, 11].

The reconstructed monocrystalline surface is an excellent seeding layer for subsequent epitaxy. We grow a first Si layer on the randomly textured surface at a temperature of 1100°C from trichlorosilane SiHCl$_3$. The layer thickness is 2 μm and the doping concentration (Boron) is $N_A = 5\times10^{18}$cm^{-3}. This epitaxial layer acts as a back surface field in the solar cell. We deposit a second epitaxial layer that is a 15-μm-thick and Boron-doped ($N_A = 1\times10^{17}$cm^{-3}) with a growth rate of 0.8 μm/min. The second layer forms the photo-active base region.

Figure 1 shows a scanning electron micrograph of a free standing detached monocrystalline thin film with an average thickness of 15 μm. The top surface is almost planar showing a pyramidal texture with facets that have a small inclination of around 8°

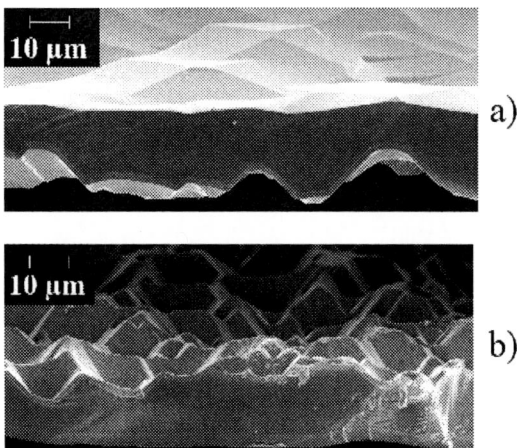

Figure 1: Scanning electron micrographs of a detached monocrystalline silicon film fabricated by the PSI process: a) a 15-µm-thick CVD-film deposited on a randomly textured substrate, b) same film from the substrate side showing random inverted pyramids.

against the macroscopic cell surface. The bottom surface shows random inverted pyramids since the substrate has random upright pyramids. The film's surface structure enables efficient light trapping. According to optical absorption measurements short circuit current densities as high as 37 mA/cm^2 are feasible with 15-µm-thick films [12].

Layer transfer and module fabrication

Free standing large area films such as those in Figure 1 can't be processed savely. Therefore our novel technique to realize an integrated series connection does support the film in every processing step either by the substrate wafer or by a glass carrier. The process is outlined in Figure 2. A photograph of the finished module is shown in Figure 3. We discuss the fabrication step by step:

Figure 2a: Prior to detaching the epitaxial film we diffuse an n-type emitter from a solid phosphorus source. For emitter contact formation, we evaporate Al through a shadow mask. The comb-like structure of the mask includes small fingers with a width of 50 µm and a busbar (see Figure 3). The module consists of 5 unit cells.

Figure 2b: In order to reduce the light reflection on the front side of the module we deposit an anti-reflection coating with a refractive index n between 2.05 and 2.1. So far the process is identical to the process we use to fabricate single cells [5].

Figure 2c: To generate an integrated series connection, we etch a 4-µm-deep and 0.5-mm-wide trench between the unit cells to separate the emitter into five regions. For this purpose, we developed a damage-free plasma etching process based on a large area microwave (MW) plasma source using Ar/SF$_6$ as reactive gas [13]. Mechanical shadow masks allow the pattern formation during the plasma process.

Figure 2d: The silicon layer is now transferred to a carrier glass. We fix the pre-processed module by a transparent glue that permits cell processing at temperatures up to 330°C to a soda-lime glass with a thickness of 1 mm.

Figure 2e: Applying mechanical stress, the separation layer breaks and the epitaxial Si film separates from the substrate. The yield for separating the 4"-wafers is above 95%. We plan to re-use the substrate wafer for the fabrication of further cells. The sintered porous silicon (SPS) layer that evolves from the porous start layer remains on the textured side of the monocrystalline Si film. We remove the SPS layer and 1 μm of the p^+-type region by plasma etching.

Figure 2f: There is no access to the embedded front side of the silicon layer after the transfer to the glass carrier. A further plasma etching step from the rear side separates the Si base material into five unit cells. Plasma etching stops at the glue and at the Al of the busbar.

Figure 2g: We evaporate Al to the backside to contact the p-type Si base. The position of the shadow mask guarantees an overlap of the back side metallization with the exposed Al busbar of the front contact. The glue protects the pn-junction from shunting. With this step, the integrated series connection is finished and we obtain a photovoltaic module that is already encapsulated. Both contacts to the module are freely accessible from the rear side.

See Figure 3 for a photo of the module. The optical shading losses due to metallization and series connection lines sum up to 19% of the total module area.

RESULTS

The fabricated module has a total area of 25 cm^2. The open circuit voltage is 3028 mV, that is an average of 606 mV for each of the 5 unit cells. The fill factor is 74.8%. The independently confirmed efficiency is 9.9% under air mass 1.5 illumination at 25°C and 1000 W/m^2. The short circuit current is 109.1 mA that is 21.8 mA per cm^2.

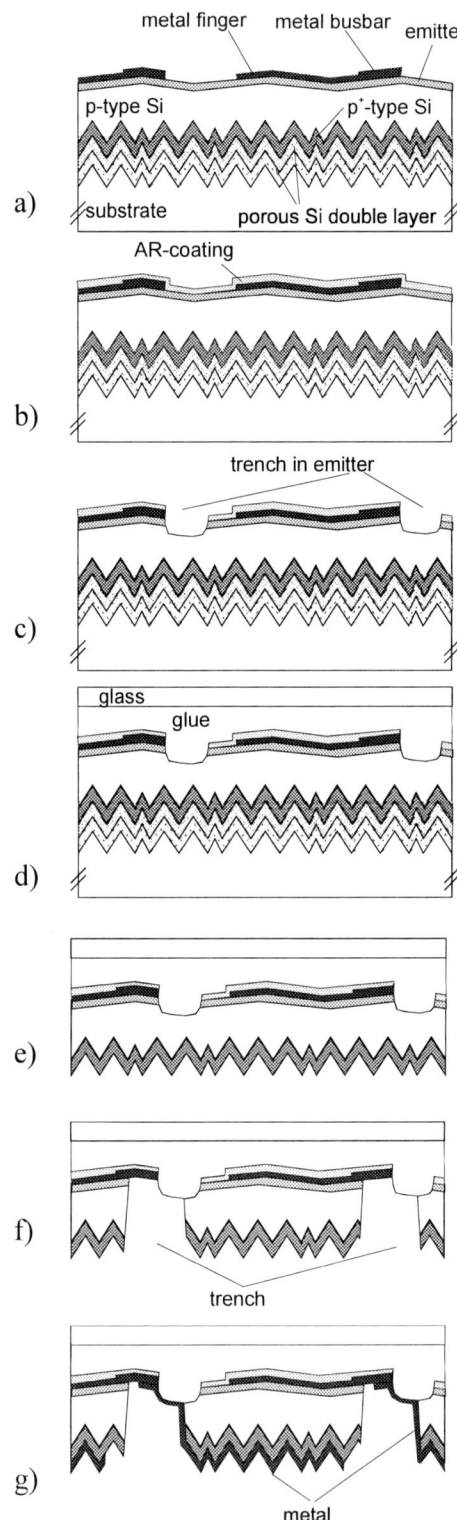

Figure 2: Processing sequence for fabricating an integrated series connection by layer transfer.

Figure 3: Photograph of a 5x5cm^2 monolithically series connected module with a confirmed efficiency of 9.9%. The Si film is 16 μm in thickness.

CONCLUSIONS

We demonstrated a novel fabrication technique for an integrated series connection for textured crystalline Si thin-film photovoltaic modules. A particular attractive feature of our process is that only one additional plasma etching step is required when compared to the process for making a single cell. The test module consists of five series connected cells. Each cell is 5×1 cm^2 in area, the module is 5×5cm^2. With a 16-μm-thick monocrystalline Si film we achieve a conversion efficiency of 9.9% and an open circuit voltage of 606 mV per cell.

Light trapping has to be improved since a short circuit current density of 21.8 mA/cm^2 is not satisfactory. If the cell had only 4% shading losses instead of 19% the expected short circuit current is 25.8 mA/cm^2. This value equals the short circuit current we measured in identically processed single cells that have 4% shading loss. A quantum efficiency analysis of these single cells reveals 5 mA/cm^2 current loss due to a poor back surface reflectance [5]. The back surface reflectance thus needs to be improved in order to really profit from the light trapping texture. We will also change the mask design to decrease optical shading losses.

The novel patented interconnection technique bases on (i) thin-film layer transfer and (ii) fast plasma etching that stops at the metal of the front contact grid. This technique should also be of interest to micro electronics. A two-layer metallization, one layer on top of the transferred film and one on the bottom, is easily achieved with the possibility to interconnect both metallization layers through plasma etched holes at arbitrary positions in the circuit.

ACKNOWLEDGMENT

The authors thank H. Artmann (Robert Bosch GmbH, Zentralbereich Forschung und Vorausentwicklung, Abt. FV/FLD) and G. Müller (ZAE Bayern) for optimizing and preparing the porous Si multilayer system, and M. Schulz (Univ. Erlangen-Nürnberg) for his continuous support of our solar cell research. The project funding by the German BMWi under contract No. 329816 is gratefully acknowledged.

REFERENCES

1. T. Yonehara, K. Sakaguchi, and N. Sato, Appl. Phys. Lett. **64**, 2108 (1994).

2. R. Brendel, in *Proc. 14th European Photovoltaic Solar Energy Conf.*, ed. by H. A. Ossenbrink, P. Helm, and H. Ehmann (Stephens, Bedford, 1997), p. 1354.

3. H. Tayanaka, K. Yamauchi, T. Matssushita, in *Proc. 2nd World Conf. Photovoltaic Solar Energy Conf.*, ed. by J. Schmid, H. A. Ossenbrink, P. Helm, H. Ehmann, E. D: Dunlop, (Joint Research Center European Comission, Ispra, 1998), p. 1272.

4. T. J. Rinke, R. B. Bergmann, and J. H. Werner, in *Proc. 16th European Photovoltaic Solar Energy Conf.*, ed. by H.Scheer, B. McNelis, W. Palz, H. A. Ossenbrink, and P. Helm (James & James, London, 2000), p. 1128.

5. R. Brendel, R. Auer, and H. Artmann, Progress in Photovoltaics (2001), in press.

6. S. Keller, S. Scheibenstock, P. Fath, G. Willeke, and E. Bucher, J. Appl. Phys. **87**, 1556 (2000).

7. H. Takato, T. Sekigawa, R. Shimokawa, in *Proc. 2nd World Conf. on Photovoltaic Solar Energy Conversion*, edited by J. Schmid, H.A. Ossenbrink, P. Helm, H. Ehmann, E. D. Dunlop, (European Commission, Ispra, 1998), p. 1810.

8. U. Kerst, B. Müller, M.E. Nell, and H.G. Wagemann, in *Technical Digest of the 11th Int. Photovoltaic Science and Engineering Conf.*, (Tanaka Printing, Kyoto, 1999), p. 747.

9. T. Matushita, S. Mizuno, H. Tayanaka, and K. Yamauchi, in *Proc. 16th European Photovoltaic Solar Energy Conf.*, ed. by. H. Scheer, B. McNellis, W. Palz, H.A. Ossenbrink, P. Helm, (James & James, London), p. 1679 (2000).

10. N. Sato, K. Sakaguchi, K. Yamagata, Y. Fujiyama, and T. Yonehara, J. Electrochem. Soc. **142**, 3116 (1995).

11. T. J. Rinke, R. B. Bergmann, R. Brüggemann, and J. H. Werner, *Ultrathin quasi-monocrystalline silicon films for electronic devices*, Solid State Phenomena **67-68**, 229 (1999).

12. G. Kuchler, D. Scholten, G. Müller, J. Krinke, R. Auer, and R. Brendel, *Fabrication of textured monocrystalline Si-films using the porous silicon (PSI)-process*, in *Proc. 16th European Photovoltaic Solar Energy Conf.*, ed. by H. Scheer, B. McNelis, W. Palz, H. A. Ossenbrink, and P. Helm (James & James, London, 2000), p. 1695.

13. R. Auer, V. Gazuz, J. Ackermann, W. Kintzel, R. Brendel, and M. Schulz, Large area plasma processes for internally series connected thin-film silicon solar cells, in *Proc. 16th European photovoltaic Solar Energy Conf.*; ed. by H. Scheer, B. McNells, W. Palz, H. A. Ossenbrink, and P. Helm; James & James: London, 2000; 1691.

Mat. Res. Soc. Symp. Proc. Vol. 681E © 2001 Materials Research Society

Transfer and handling of thin semiconductor materials by a combination of wafer bonding and controlled crack propagation

J. Bagdahn[1,2], D. Katzer[1], M. Petzold[1], M. Wiemer[3], M. Alexe[4], V. Dragoi[4], U. Goesele[4]

[1] Fraunhofer Institute for Mechanics of Materials, Heideallee 19, D-06120 Halle, Germany.

[2] Johns Hopkins University, Department of Mechanical Engineering, 3400 N. Charles Street, Baltimore MD 21218-2681, U.S.A.

[3] Fraunhofer Institute for Reliability and Microintegration, Dept. Micro Devices and Equipment, Postfach 344, D-09003 Chemnitz, Germany

[4] Max- Planck-Institute of Microstructure Physics, Weinbergweg 2, D-06120 Halle, Germany.

ABSTRACT

Direct waferbonding is an appropriate technology to join two or more wafers of the same or of different materials. Waferbonding can be used to stiffen thin wafers during fabrication. However, conventional fabrication processes lead to an increase of the bond strength, which inhibits the required de-bonding. The propagation of cracks, which is based on a subcritical crack growth in the bonded interface, was used to cleave the bonded wafers. The subcritical crack growth is limited to the bonded interface, since the adjacent bulk semiconductor materials are inherently resistant to subcritical crack growth. The process allows the separation of Si-Si and Si-GaAs wafers after annealing. Wafer-bonded SOI wafers can also be separated with this technology even if they were annealed at 1100°C. The first examples for wafer stiffening during fabrication and wafer transfer using the developed approach will be presented.

INTRODUCTION

The fabrication of microelectronic, micromechanical and optoelectronic devices requires manufacturing, handling or transferring of thin layers of semiconductor materials with thickness of 250 μm or less. Direct waferbonding is an appropriate technology to join two or more wafers of the same or of different materials /1, 2/. Therefore the bonding of a thin process wafer to a thick substrate wafer can be used to stiffen the thin process wafer and avoid wafer bowing or fracture. In addition, the approach enables the use of conventional high temperature processes such as oxidation, diffusion, or film deposition. However, these steps contribute to a strength increase in the bonded interface and would prevent the required de-bonding, of the process wafer from the handling substrate wafer, without wafer fracture. Therefore, a reliable separation technique for high-strength bonded wafers can be of practical significance.

In this paper, we will present a new method /3/, which stiffens thin semiconductor materials during fabrication by direct waferbonding and subsequently promotes controlled cleaving by using subcritical crack growth in the bonded interface. It is the aim of the paper to demonstrate the application of this technique in combination with wafer bonding for the handling of GaAs and the fabrication of Si microelectromechanical systems (MEMS) using SOI wafers.

CLEAVING APPROACH

It was recently shown that subcritical crack growth processes can occur in the interface of mechanically loaded, wafer-bonded semiconductor materials /4/. The subcritical crack growth determines the lifetime of stressed wafer-bonded MEMS and, therefore, their mechanical reliability properties. For instance, it was shown that a wafer-bonded device, which is statically loaded with 30-40% of its initial strength, fails after a time of about 1-½ years /5/. The subcritical crack growth is based on the stress corrosion of siloxane bonds (Si-O-Si) in the bonded interface. Molecules which contribute to the stress corrosion processes of siloxane bonds are e.g. water, ammonia, or methanol /6/. For example, the humidity of the ambient air is a typical source of water molecules. Figure 1 shows the subcritical crack growth in a silicon dioxide network. In the first step the water molecules will be readily transported to the crack tip (Figure 1, step 1). Afterwards, the water molecules attack highly stressed siloxane bonds at the crack tip (Figure 1, step 23). As a result, silanol bonds (Si-OH) were formed as follows:

$$Si\text{-}O\text{-}Si \ + \ H_2O \qquad \xrightarrow{\text{mechanical stress}} \qquad Si\text{-}OH : HO\text{-}Si \qquad (1)$$

The concomitant hydrogen bonds between opposing silanol groups can be more easily cleaved by the applied stress compared to the covalent Si-O bonds (Figure 1, step 3). After cleaving one siloxane bond, the process will continue at the next siloxane bond. The velocity of rupture depends on the applied mechanical load and the amount of reactive species at the crack tip. Therefore the intensity of the applied load and the environmental conditions influence the velocity of crack propagation.

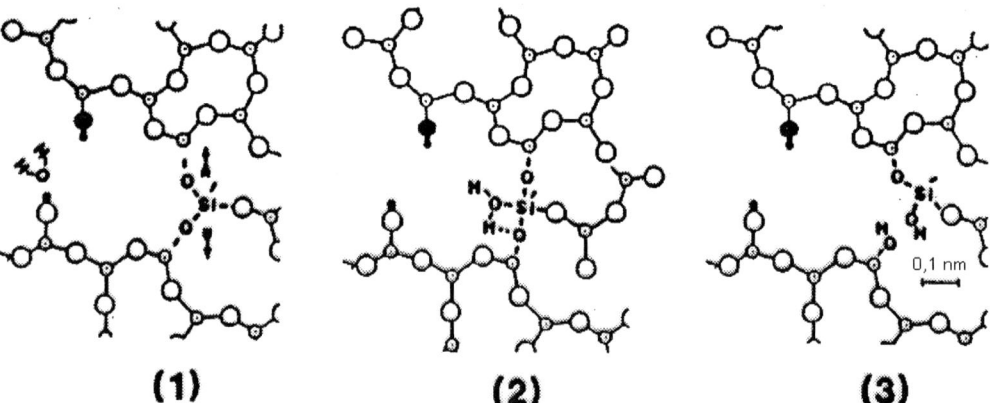

Figure 1: Stress corrosion in a silicon dioxide network (after Michalske /6/), see text for details.

In Figure 2 the subcritical (slow) crack growth velocity of SOI (silicon-on-insulator) wafers with different annealing temperatures is depicted. The crack growth velocity, v, in Figure 2 is plotted as a function of the relative loading ratio, R, where R is the ratio between the applied loading and the inertial strength of the interface. A value of R=1 leads to a fast continuous crack growth, which is frequently combined with an out-kinking of the crack into the bulk material. It can be seen in Figure 2 that the maximum subcritical crack growth velocity can reach values up to $v \sim 10^{-5}$–10^{-4} m/s for samples annealed at high temperature, when they are cleaved under laboratory conditions. However, the subcritical crack growth velocity can be further increased by changing the environmental conditions, e.g. the cleaving in pure water leads to an increase in the maximum velocity up to $v = 10^{-3}$ m/s /5/.

As a consequence, slow crack propagation with a velocity in the order of a few nanometers up to millimeters per second can be observed depending on the applied stress level and the environmental conditions. The investigations revealed that subcritical crack growth occurs in bonded interfaces that contain a homogeneous silicon dioxide layer (siloxane bonds). It was found that hydrophobic bonded interfaces and hydrophilic interfaces with a native – native oxide annealed at 1100°C showed no subcritical crack growth behavior. This is reasonable since siloxane bonds are absent in a hydrophobic interface and local regions of silicon-silicon bonds interrupt the native silicon dioxide layer in hydrophilic samples annealed at 1100°C /7/. Since the silicon-silicon bonds are inherently resistant to stress corrosion, the slow crack growth in these bonded interface is prevented.

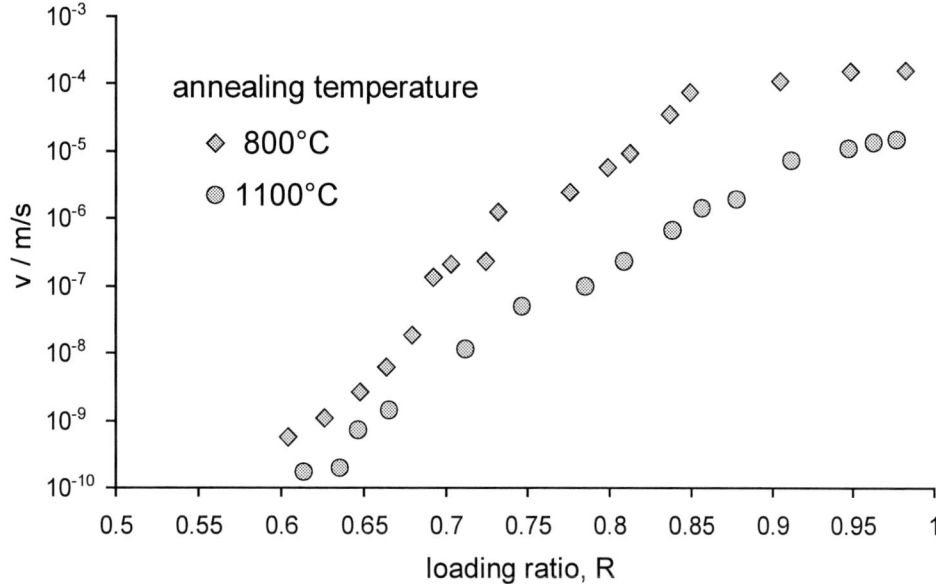

Figure 2: Subcritical crack growth velocity, v, versus loading ratio R. Samples: Directly bonded 4'' Cz-silicon wafer (thickness 525 μm) with a 500 nm interfacial oxide. Environmental conditions during cleaving: 23°C and 30% relative humidity.

In contrast to the bonded interface, most of the semiconductor bulk materials, such as Si or GaAs, are known to be insensitive to crack corrosion. Therefore, the subcritical crack growth is confined to the wafer bond interface preventing the crack from kinking into the wafer material. The effect enables controlled crack propagation in the bonded interface and may, consequently also be utilized for the cleaving of bonded wafers.

The developed handling and transferring approach consists of three basic steps. First, a process wafer is bonded to a handling wafer. Subsequently, the process wafer can be grinded, etched or polished to the required thickness. In the second step, all necessary technological processes such as oxidization, implantation, film deposition, high temperature annealing or etching can be performed on the process wafer without any restrictions due to the handling substrate. In the third step, the process wafer is cleaved from the handling wafer using a controlled subcritical crack growth in the bonded interface. The handling wafer is not destroyed and can be used again for future processing steps. For the cleaving step, a special computer controlled device was developed that inserts thin blades at four different positions into the bonded interface. During the cleaving process, the four blades will be inserted into the bonded interface and fatigue cracks will form in front of the blade tips. In order to generate the subcritical crack growth, a controlled slow insertion speed is required during the initial stage. The cracks grow in front of the blades until they are united, which leads to the complete separation of the bonded wafers. A computer controlled cleaving device was used for the regulation of the insertion velocity. If the initial cracks are formed, the crack growth velocity can be increased to a few hundred μm/s up to few mm/s. The maximum cleaving velocity depends on the annealing treatment, the ratio of thickness between the wafers, and the environmental conditions during cleaving. Details of the technical approach can be found in Bagdahn et al. /8/. In Figure 3, a completely cleaved SOI wafer (annealed at 1100°C) is presented. An interference pattern is visible on the cleaved surfaces, due to the fracture surface roughness in the nanometer range, caused by the crack growth in the 500 nm thick interfacial oxide.

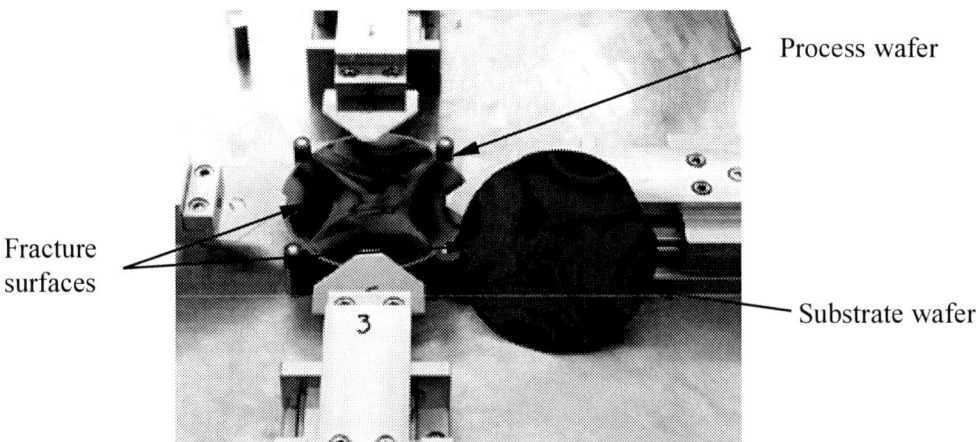

Figure 3: Cleaved bonded 4" (100) SOI wafer pair after an annealing of 1100°C.

HANDLING OF GaAs WAFERS DURING FABRICATION

It was recently shown that intermediate spin on glass films can be used for bonding of Si-GaAs wafers at low temperature /9/, which enables the fabrication of Si-GaAs heterostructures. Furthermore, the bonding technology can be used for strengthening of brittle GaAs wafers. Si-Si and Si-GaAs were bonded and subsequently annealed at 200°C in order to study the subcritical crack growth behavior in the intermediate glass layer. The investigation revealed a slow crack propagation in the intermediate glass layer. Figure 4 shows the crack growth velocity in the intermediate glass layer as a function of the loading ratio, R, between Si-Si and Si-GaAs wafers, which were annealed at 200°C. The results in Figure 4 show that for the same bonding conditions the crack growth velocity depends on the material combination. This effect can be attributed to the stress situation at the crack tip. The crack is loaded by pure tensile stresses during cleaving of the Si-Si wafers, whereas the difference in the elastic material properties between Si and GaAs lead to additional shear forces and a reduction of the tensile force at the crack tip. This reduces the crack growth velocity since the intensity of the tensile force is responsible for the bond rupture on the microscopy scale /10/.

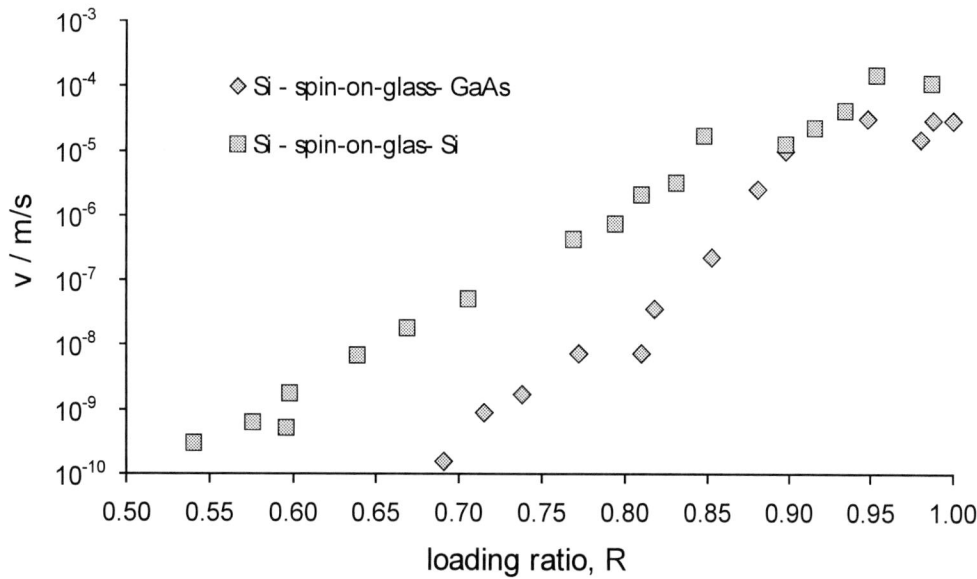

Figure 4: Subcritical crack growth velocity, v, versus loading ratio, R, for spin on glass bonded wafers. Samples: 4'' Cz-silicon wafer (thickness 525 μm) and 4'' Cz-silicon wafer (thickness 525 μm) – GaAs wafers (thickness 650 μm). Environmental conditions during cleaving: 23°C and 30% relative humidity.

FABRICATION OF MICROMECHANICAL SYSTEMS

Figure 5 shows the process flow during fabrication of a MEMS device by transferring a thin wafer to another wafer.

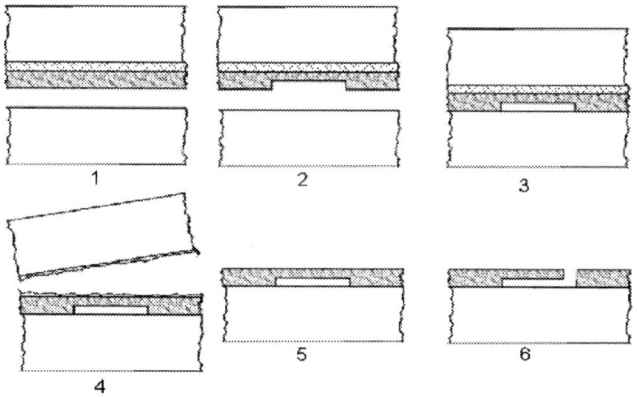

Figure 5: Transferring thin semiconductor materials by direct waferbonding and controlled cleaving to form a micromechanical device (see text for details).

A patterned SOI wafer is bonded against a flat silicon wafer (Figure 5-1-3). Afterwards, the SOI wafer is cleaved at the high temperature bonded interface using the described approach (Figure 5-4). The fabricated new wafer contains a sealed cavity, which can be used for pressure sensors or fluidic devices (Figure 5-5). Further etching processes can be applied to produce movable elements (Figure 5-6), therefore a freestanding beam with a defined thickness and a possible large displacement perpendicular to the bonded interface is produced. The cleaved wafer from the SOI wafer can be used either as a new handling or as a capping wafer. The investigations showed the possibility of transferring a 50 µm thick thin silicon wafer from a wafer-bonded SOI wafer (annealing 1100°C, 500 nm interfacial oxide) to a second wafer by cleaving the high temperature bond interface even if the second bond was only annealed at 400°C.

CONCLUSIONS

We have shown that the propagation of subcritical cracks can be used to separate directly bonded Si-Si and Si-GaAs wafers. The controlled cleaving in the bonded interface is based on a stress corrosion process in the bonded interface. The velocity of the crack growth varied between a few nanometers to millimeters per second, depending on the bonding conditions, the environmental conditions during cleaving and the thickness ratio between the separated wafer levels. The technology can be used for stiffening of thin and brittle wafers during fabrication, even at high temperatures. Furthermore, the approach supports the fabrication of microelectromechanical systems, further studies about the fabrication of a pressures sensor are under investigations.

REFERENCES

/1/   Q.-T. Tong and U. Goesele, Semiconductor wafer bonding: Science and Technology, New York: John Wiley & Sons. Inc. (1999).

/2/   A. Ploessl and G. Kraeuter, Wafer direct bonding: Tailoring adhesion between brittle materials, Materials Science & Engineering Report (R25). Elsevier Amsterdam, Lausanne, New York, Oxford, Shannon, Tokyo (1999).

/3/   J. Bagdahn and M. Petzold, Handling and transferring of thin semiconductor materials, Patent pending.

/4/   J. Bagdahn, D. Katzer and M. Petzold, Investigations of the subcritical crack growth in wafer-bonded interfaces (in German), DVM-Band 230 "Unterkritisches Risswachstum": 385-394, 1998.

/5/   J. Bagdahn and M. Petzold, Fatigue of directly wafer-bonded silicon under static and cyclic loading, Journal of Micro System Technology, in press.

/6/   T.A. Michalske, Fundamental studies of glass fracture, Proc. XV Int. Congress on Glass, 3-15, 1989.

/7/   J. Bagdahn and M. Petzold, Lifetime properties of wafer-bonded components under static and cyclic loading, Fifth Int. Symp. on Semicond. Wafer Bonding: Science, Technol. and Appl., 1999, Honolulu, Hawaii, in press.

/8/   J. Bagdahn, D. Katzer, M. Petzold, M. Wiemer, M. Alexe, V. Dragoi, U. Goesele, A new approach for handling and transferring of thin semiconductor materials, Proceed. Micro System Technologies 2001, H. Reichl (ed.), VDE Verlag Berlin u. Offenbach 2001, S. 449-454

/9/   M. Alexe, V. Dragoi, M. Reiche and U. Goesele, Low temperature GaAs/Si wafer bonding. Electronics Letters, Vol. 36, No. 7, 677-678.

/10/   B.R. Lawn, Fracture of brittle solids, Cambridge university press, 2nd edition, 1995.

Mat. Res. Soc. Symp. Proc. Vol. 681E © 2001 Materials Research Society

Gettering Control at Bonding Interface in ELTRAN®

Kazutaka Momoi, Masataka Ito, Nobuhiko Sato, Noriaki Honma and Takao Yonehara
ELTRAN Business Center, Canon Inc.,
6770 Tamura, Hiratsuka, Kanagawa, 254-0013, Japan.

ABSTRUCT

Heavy metal gettering capability on ELTRAN® was studied by controlling surface treatments for handle wafer prior to wafer bonding. Hydrophobic bonding pre-treatments wafer had much higher heavy metal gettering capability at bonding interface than hydrophilic wafer. In the case of hydrophilic bonding pre-treatments, atomically flat bonding interface was observed by cross-sectional TEM. On the other hand, in the case of hydrophobic bonding pre-treatments, "nano gaps" were observed at bonding interface. We concluded that these differences in the structure at the bonding interface caused the difference in the gettering capability. It is possible to control gettering capability by no additional steps in SOI wafer process.

INTRODUCTION

When an SOI (Silicon on Insulator) wafer is contaminated by heavy metals in device fabrication process, gettering layer in SOI wafer can reduce these contaminations from device active layer. It is one of the keys to control the gettering capability for realization of robust SOI device fabrication process.

In resent studies on heavy metal gettering in bonded SOI wafer, K. Okonogi has reported that Cu was not captured at bonding interface after annealing at 900 °C for 2 hours [1]. He has concluded that the bonding interface between thermal SiO_2 and single crystal Si wafer was not an effective gettering site. On the other hand, J. Furihata et. al. have reported that Cu was captured at bonding interface after annealing at 700 or 900 °C for 1 hours in N_2 + 2 % O_2 ambient [2]. These contradictory results suggest that the metal gettering capability at bonding interface has not been controlled in these studies.

In this report, the correlation of the metal gettering capability at the bonding interface with handle wafer treatments prior to wafer bonding was studied for ELTRAN® SOI-Epi wafer™, which is characterized by bonding of epitaxial Si on porous Si, splitting and etching back of porous Si [3, 4].

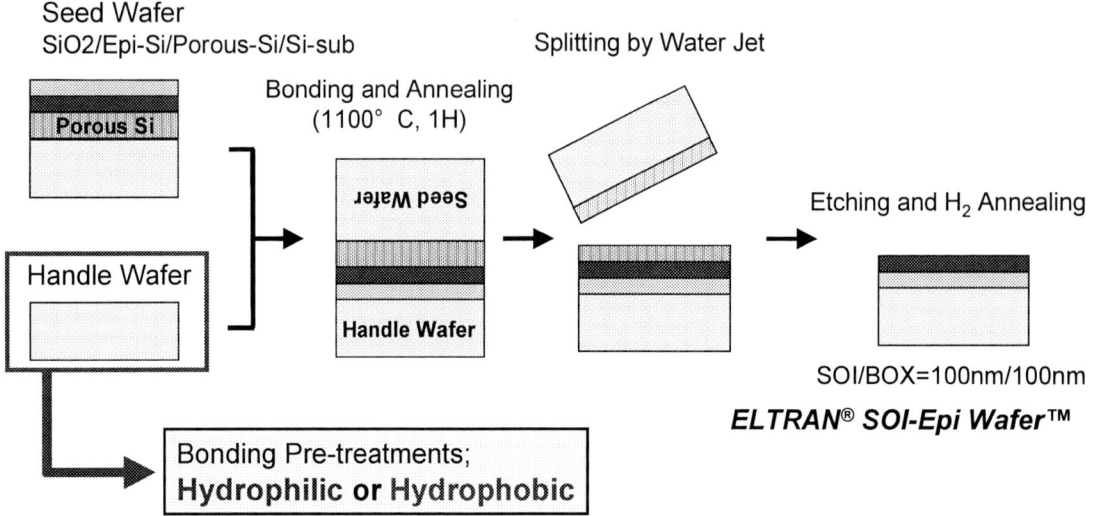

Figure 1. ELTRAN process and experimental conditions.

EXPERIMENTAL DETAILS

ELTRAN process and the experimental conditions are shown in Figure 1. A seed wafer which has SiO_2/Si/porous-Si/Si-sub structure was bonded with a handle wafer. Both hydrophobic and hydrophilic handle wafer surfaces were examined for the comparison of heavy metal gettering capability. Following wafer bonding, these bonded wafers were annealed at 1100 °C for 1 hour for bond strengthening. After splitting and etching back of porous Si, the surfaces in these wafers were annealed for surface flattening. The experimental ELTRAN wafers had the thickness of SOI layer = 100 nm and BOX layer = 100 nm.

In order to compare the gettering capability at bonding interface, the surfaces of experimental wafers were intentionally contaminated by Fe, Ni and Cu standard aqueous solutions by using micro drop method [5]. These solutions were dropped at about 100 points onto the wafer surface, and then dried up on a hot plate. The initial concentration on the wafer surface was controlled at 6E+12 atoms/cm^2 on each element. These ELTRAN wafers were annealed at 1050 °C for 3 hours in N_2 ambient to diffuse metals into the wafer. After intentional contamination and annealing, the gettering capability at bonding interface was evaluated with SIMS (Secondary Ion Mass Spectroscopy, CAMECA IMS-4f) and ICP-MS (Inductively Coupled Plasma Mass Spectroscopy, Seiko Instruments SPQ9000). In the SIMS analysis, O_2^+ ion, which is accelerated to 8.0 kV, was used for sputtering on the sample. On the other hand, the experimental ELTRAN wafers were chemically etched step-by-step by HF or HF + HNO_3 mixed acid for ICP-MS analysis.

To investigate detail structure of the bonding interface, cross-sectional images were taken by using Hitachi H-9000NAR.

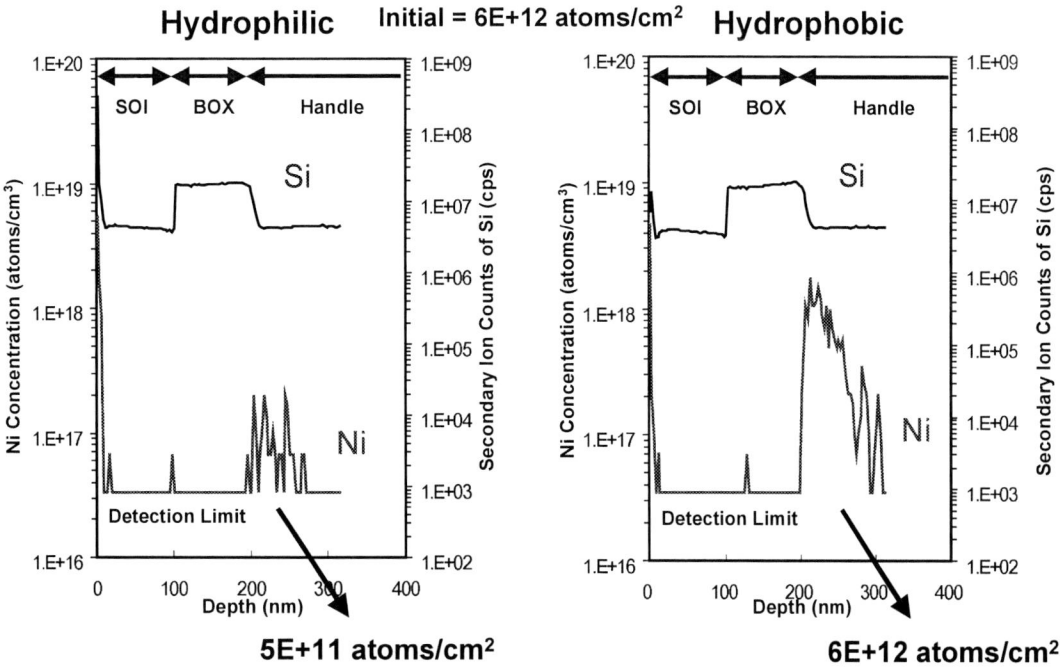

Figure 2. Comparison of SIMS depth profiles between hydrophilic and hydrophobic bonding pre-treatments on intentionally contaminated (Ni) and annealed ELTRAN wafers.

RESULT AND DISCUSSION

Figure 2 shows the comparison of SIMS profiles between hydrophilic and hydrophobic bonding pre-treatments on intentionally contaminated by Ni and annealed ELTRAN wafers. Blue line shows secondary ion counts of Si, and red line shows Ni concentration. Ni concentration at the bonding interface between BOX layer and handle wafer in hydrophilic bonding pre-treatments was about 5E+11 (atoms/cm^2). This concentration is less than one in ten compared to initial concentration. On the other hand, 6E+12 (atoms/cm^2) of Ni was captured in hydrophilic bonding pre-treatment wafer. In other wards, this result means that almost initial Ni atoms were captured at bonding interface in hydrophobic wafer. These results indicate that the bonding interface in hydrophobic wafer has much higher gettering capability than in hydrophilic wafer.

Figure 3 shows results of ICP-MS analysis on the metal capture capabilities at bonding interface. Metal concentration at bonding interface in hydrophilic bonding pre-treatments wafer is assumed as unity on each metal for the comparison of gettering capability. The hydrophobic bonding pre-treatments wafer had from 3 (Cu) to 12 (Ni) times higher metal capture capabilities than hydrophilic pre-treatments wafer. This result indicates that gettering capability in ELTRAN wafer on Fe, Ni and Cu was effectively controlled by bonding pre-treatments with no additional steps for the formation of gettering layer.

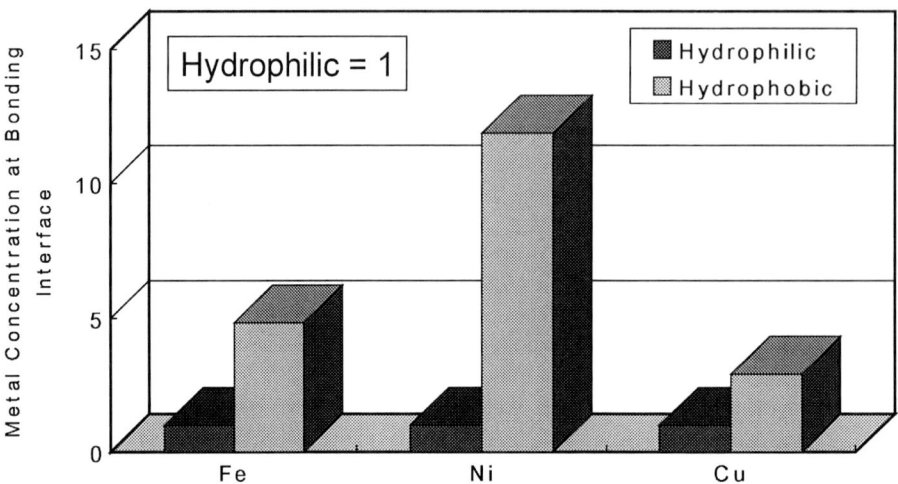

Figure 3. Comparison of heavy metal concentration at bonding interface between bonding pre-treatments analyzed by ICP-MS in conjunction with chemical step etching method.

In order to investigate the difference of the gettering capability, the bonding interface in ELTRAN wafers were observed by cross-sectional TEM (Figure 4). The bonding interface in hydrophilic bonding pre-treatments wafer was as smooth as the interface between SOI and BOX layer (figure 4(a) top). Even in an atomic image (figure 4(a) bottom), the interface was smooth. A similar atomically flat bonding interface was observed by Y. Kawai et. al. [6].

On the other hand, the characteristic small cavities, named "nano gaps", were observed at the bonding interface in a wafer of hydrophobic bonding pre-treatments (figure 4(b) top). It was shown from a higher magnification image (figure 4(b) bottom) that the nano gap was composed of a cavity and inner walls which are faceted to (111) or (331) plane and covered with thin insulator (SiO_x) film.

During high temperature annealing in this experiments, heavy metals diffuse uniformly in SOI wafer. With the decreasing temperature, heavy metals around surface layer are gradually condensed locally at metal capture sites. Wafer surface is normally known one of strong metal capture sites by the existence of dangling bonds or surface defects. But in the case of a wafer with nano gaps, because they work as inner surface, heavy metals which are in handle wafer are prevented to diffuse to SOI or BOX layer and captured at nano gaps.

We speculate that *nano gaps* are formed by local thermal etching and oxidation. Adsorbed H_2O clusters from air and/or residual water at cleaning process are put between bonded wafers. The interactions of H_2O with Si under an experimental condition are expressed by next three reactions [7].

$$Si\,(s) + H_2O \leftrightarrow SiO\,(g) + H_2\,(g) \qquad (1)$$
$$Si\,(s) + 2H_2O \leftrightarrow SiO_2\,(s) + 2H_2\,(g) \qquad (2)$$
$$Si\,(s) + SiO_2\,(s) \leftrightarrow 2SiO\,(g) \qquad (3)$$

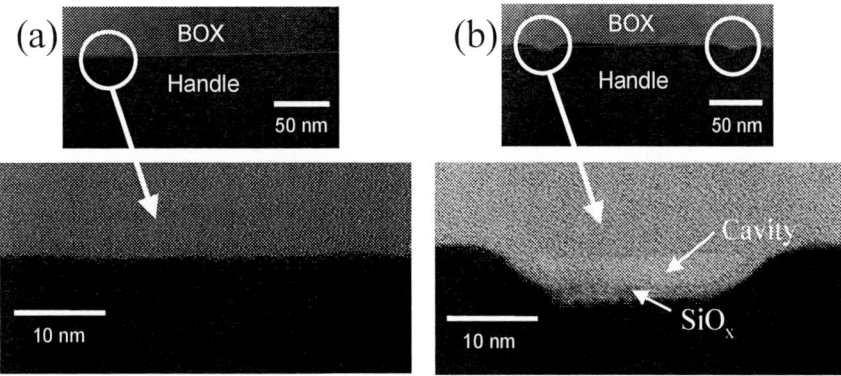

Figure 4. Cross-sectional TEM images at the bonding interface in ELTRAN wafers. An atomically flat bonding interface is observed in picture (a) in the hydrophilic treatments. On the other hand, "nano gaps" are observed in picture (b) in the hydrophobic treatments.

At high temperature in bonding annealing, reaction (1) and (2) proceed to right direction, and generate SiO(g), SiO_2(s) and H_2(g). When the reaction (1) proceeds to right direction, the handle wafer surface is etched locally, and then *nano gaps* are formed at the bonding interface.

With the decreasing temperature, reaction (3) proceeds to left for more steady state. Therefore, the walls of *nano g*ap are covered with insulator (for example SiO_2). But more detail mechanism is now under consideration.

In contrast to hydrophobic bonding pre-treatments, uniform H_2O layer exists on a handle wafer surface in hydrophilic pre-treatments. The H_2O layer is put between bonded wafers. At high temperature in bonding annealing, the reaction (1) and (2) proceed uniformly on entire handle wafer surface at the bonding interface. This indicates that local *nano gaps* are not formed in the case of hydrophilic bonding pre-treatments wafer.

CONCLUSION

Heavy metal gettering capability on ELTRAN® SOI-Epi wafer was studied by controlling surface treatments prior to wafer bonding. A hydrophobic bonding pre-treatments wafer had much higher the metal capture capability than a hydrophilic one. It was possible to control the gettering capability by the surface treatments prior to wafer bonding. A flat bonding interface was observed in the case of hydrophobic pre-treatments wafer by TEM. On the other hand, *nano gaps* were observed at bonding interface in the case of hydrophobic treatments. This technology would be promising with the aspect of cost and time, because no additional steps for the formation of gettering layer are necessary for SOI wafer process.

REFERENCES

1. K. Okonogi, *Science of Silicon,* ed. UCS Semiconductor Technology Symposium (Realize Inc., Japan, 1996), p. 609

2. J. Furihata, M. Nakano and K. Mitani, Jpn. J. Appl. Phys. **39**, 2251 (2000).

3. T. Yonehara, K. Sakaguchi and N. Sato, Appl. Phys. Lett. 64, 2108 (1994).

4. T. Yonehara, K. Sakaguchi and N. Sato, Proc. 9th Int. Symp. on Silicon-on-Insulator Tech. And Devices, 99-3, The Electrochemical Society, Seattle, p. 111 (1999).

5. H. Kondo, J. Ryuta, E. Morita, T. Yoshimi and Y. Shimanuki, Jpn. J. Appl. Phys. **31**, L11 (1992).

6. Y. Kawai, S. Ishigami, H. Furuya, T. Shingyouji and Y. Saitoh, PROCEEDINGS OF THE SECOND INTERNATIONAL SYMPOSIUM ON SEMICONDUCTOR WAFER BONDING: SCIENCE, TECHNOLOGY, AND APPLICATIONS, Hawaii, p. 216 (1992).

7. G. Ghidini and F. W. Smith, J. Electrochem. Soc. **131**, 2924 (1984).

Mat. Res. Soc. Symp. Proc. Vol. 681E © 2001 Materials Research Society

Characterization of Optical Lifetime in Silicon-on-Insulator Wafers by Photoluminescence Decay Method

Shigeo Ibuka, Michio Tajima and Atsushi Ogura[1]
Institute of Space and Astronautical Science, Sagamihara, 229-8510, Japan.
[1]System Devices and Fundamental Research, NEC Corporation, Tsukuba, 305-8501, Japan.

ABSTRACT

We report observation of temporal decay of luminescence due to electron-hole condensation in silicon-on-insulator (SOI) wafers. The condensate luminescence was observable in SOI wafers under ultraviolet light excitation, because of shallow penetration depth of the light and confinement of photo-excited carriers in the top-Si layer. We found that the temporal decay of the luminescence depended on the surface/interface condition and fabrication method. These findings can be explained by the difference in the recombination process via surface, interface and defect states in the top-Si layer. We propose that the decay measurement of the condensate luminescence has great potential for characterization of SOI wafers.

INTRODUCTION

The silicon-on-insulator (SOI) wafer is one of the most attractive materials for fabrication of next generation devices with superior properties of low-power, high-speed and radiation-hardness. With increasing application of these wafers, there is strong and growing requirement for more sensitive and non-destructive characterization of the top-Si layer with a thickness of less than 200 nm. We previously reported that photoluminescence (PL) using an ultraviolet (UV) laser light as an excitation source has been an effective tool for characterization of SOI wafers[1,2]. The UV light excitation enables us to detect the luminescence from the top-Si layer, because of its shallow penetration depth and presence of a buried oxide (BOX) layer acting as a diffusion barrier of photo-excited carriers. As a result, information about type and distribution of defects in the top-Si layer is obtained from the PL spectrum and mapping[2]. Furthermore, luminescence due to the condensed phase of high-dense carriers, known as electron-hole droplets (EHD), is detectable under the UV light excitation. The detection can be explained by confinement of photo-excited carrier in the top-Si layer[3,4].

In this paper, we studied temporal decay of the EHD luminescence in a wide variety of commercial SOI wafers[5]. The observation of the decay was realized employing a pulsed UV light as an excitation source. The decay of SOI wafers with a surface oxide was slower than that of the wafers without the oxide for all wafers. The decay of a bonded SOI wafer with a bonding interface between the top-Si and BOX was faster than that of the wafer with the interface between the BOX and substrate. These results indicate that the recombination via surface and interface states is effective for the decay process of the EHD luminescence. There was dependence of the temporal decay on the fabrication techniques, although there were negligible differences in the luminescence spectral shape among the wafers. We will demonstrate that the

195

decay was sensitive to crystalline quality of the top-Si layer, suggesting that the present method is a powerful tool for characterization of SOI wafers.

EXPERIMENTAL DETAILS

Samples used in this study were prepared from two kinds of commercial SOI wafers, separation by implanted oxygen (SIMOX) and bonded SOI. The wafers were obtained from different companies. The SIMOX wafers were fabricated under different conditions of the implanted dose, and were distinguished as SX-1 and SX-2. The bonded SOI wafers had a thin top-Si layer achieved by different techniques, and were distinguished as BS-1 and BS-2. For BS-2, two wafers with different bonding interfaces were prepared: One had a bonding interface between a top-Si and BOX layer, and the other had it between a BOX and substrate. Thickness of the top-Si layer of all the samples was 170-200 nm. Two samples were cut out from each wafer, and subjected to common treatments to equalize surface conditions. One was etched with an HF solution to remove the surface oxide layer. The other was annealed in oxygen atmosphere for formation of surface oxide layer with a thickness of 50 nm after the HF etching.

The samples were immersed in liquid He in a glass cryostat. Spectral measurement was performed using UV light from a continuous-wave Ar^+ laser operated at 351, 364 nm with a beam diameter of 0.1 mm. PL was analyzed with a grating monochromator and detected by a Ge pin diode (Northcoast EO-817L). Spectral response of the measurement system was calibrated with blackbody radiation. Time-resolved measurement was performed using a pulsed N_2 laser operated at 337 nm with a beam diameter of 1 mm. The pulse width was 3 nsec and the repetition frequency 20 Hz. Average power at 20 Hz was 3 mW on the sample surface. PL was analyzed by the same system as the spectral measurement and detected by photomultiplier (Hamamatsu R5509-41). The signal was averaged by a digital oscilloscope with a 500 MHz bandwidth. The penetration depth of the two UV laser lights for a Si crystal was estimated as about 10 nm[6].

RESULTS AND DISCUSSION

Figure 1 shows PL spectra of commercial SOI wafers at 4.2K under UV light excitation. Units of the vertical axis are commonly scaled for all the spectra and the symbol "x50" denotes the relative amplitude factor. The subscript "TO" in the label denotes the transverse-optical-phonon sideband. In all the wafers, the luminescence line due to recombination of carriers in the EHD was predominantly observed at the energy of around 1.09 eV, and deep-level luminescence was negligible. There was no difference of the spectral shape among the SOI wafers, and it was substantially the same as that of as-grown bulk Si under high intensity excitation. This result implies that the spectral measurement is not very effective for characterization of commercial SOI wafers.

The temporal decay of the EHD luminescence of BS-1 at 4.2K is shown in Fig. 2. The EHD luminescence (EHD-PL) lifetime τ_{EHD-PL} is determined from a temporal decay of the

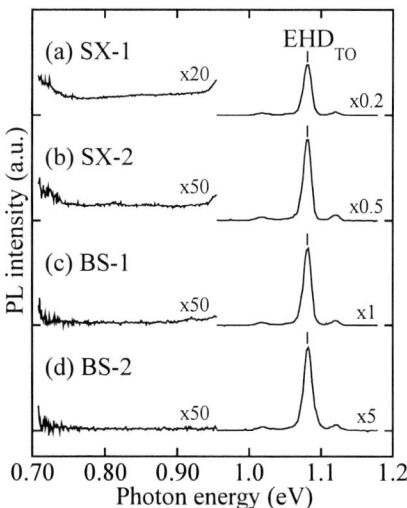

Figure 1. Comparison of PL spectra of commercial SOI wafers at 4.2 K under continuous-wave UV light excitation.

luminescence,

$$I_{EHD-PL} \propto \exp(-t / \tau_{EHD-PL}), \qquad (1)$$

where I_{EHD-PL} is intensity of the luminescence. If the EHD recombination is an intrinsic process, the EHD-PL lifetime may be described as,

$$1/ \tau_{EHD-PL} = 1/ \tau_{EHD} + 1/ \tau_{NR}, \qquad (2)$$

where τ_{EHD} is the intrinsic lifetime of the EHD, and τ_{NR} is non-radiative recombination lifetime. The EHD-PL lifetime was calculated from the time constant of an initial exponential decay.

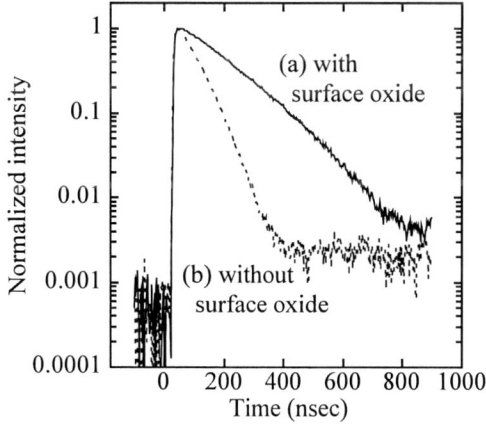

Figure 2. Temporal decay of the EHD luminescence of sample BS-1: (a) with surface oxide, (b) without surface oxide.

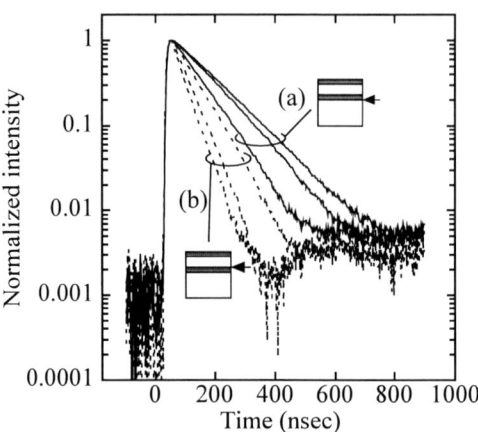

Figure 3. Dependence of temporal decay of the EHD luminescence on the bonding position in sample BS-2: (a) bonded at the BOX/substrate interface, (b) bonded at the top-Si/BOX interface. The arrow in the inset indicates the bonding position.

The latter part had a slightly faster decay than the initial part. This can be explained by the fact that the EHD with a size smaller than the critical size disappears rapidly[7]. In the wafer without a surface oxide, the decay was very fast and the EHD-PL lifetime was estimated as 68 nsec. When the surface oxide was formed, the decay slowed with the lifetime of 157 nsec; this value is comparable to that of a bulk Si[8]. The increase of the EHD-PL lifetime after formation of the oxide layer was observed in all the SOI wafers. We suggest that the quick decay of the wafers without this layer originates from the fast recombination velocity via surface states.

Figure 3 shows dependence of the temporal decay of the EHD luminescence on a bonding position in BS-2 at 4.2 K. The arrows in the inset illustration indicate the bonding position. The measured decay obviously varied with the measured position on both BS-2 wafers. We recognized a variation of intensity in the spectral measurement, which was attributable to inhomogeneity of crystalline damage induced during the thinning process of the top-Si layer. We, therefore, measured the decay at three points in each wafer, as shown in this figure. All decays of the wafer bonded at the top-Si/BOX interface were faster than those of the wafer bonded at the BOX/substrate interface, indicating that recombination velocity of the bonding interface is faster than that of the thermally-oxidized interface. These results allow us to propose that observation of the EHD-PL lifetime is applicable for characterization of surface and bonding interface.

Comparison of PL decay curves of the EHD luminescence among the commercial SOI wafers with a surface oxide layer at 4.2 K is shown in Fig. 4. The EHD-PL lifetime of SX-1, SX-2, BS-1 and BS-2 (bonded at the BOX/Substrate interface) are estimated as 150, 114, 158 and 88 nsec[9], respectively. The EHD-PL lifetime of SX-2 and BS-2 was obviously shorter than that of SX-1 and BS-1, which were comparable to that of a bulk Si. The wafers had the same surface condition, since they were subjected to the same treatment of HF etching and oxidation. We believe that the fast decay of the EHD luminescence originates from poor

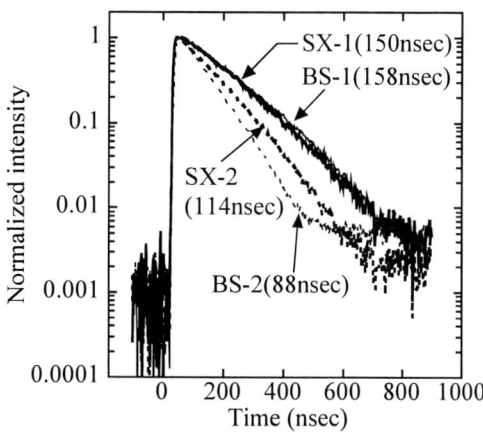

Figure 4. Comparison of temporal decay of the EHD luminescence in various SOI wafers. The data of sample BS-2 is the fastest decay data shown in Fig. 3(a). The EHD-PL lifetime of each wafer is shown in parentheses.

quality of the top-Si and interface between the top-Si and BOX layer. In the two bonded SOI wafers, the top-Si/BOX interface was formed by the thermal oxidation. Therefore, the difference of the EHD-PL lifetime between BS-1 and BS-2 is attributable only to the crystalline quality of the top-Si layer after different thinning processes. We propose that the decay measurement of the EHD luminescence is applicable to characterization not only of the surface and interface states, but also of the crystalline quality of the top-Si layer.

The mechanism of variation of the EHD-PL lifetime associated with surface, interface and defect states is under investigation. Based on time-resolved spectra and temperature dependence of the decay, we speculate that the EHD-PL lifetime depends on the concentration of carriers confined in the top-Si layer immediately after the optical excitation. Detailed discussion on this will be reported in the near future in a separate paper.

CONCLUSIONS

We observed temporal decay of the EHD luminescence in various SOI wafers under pulsed UV light excitation. The short EHD-PL lifetime was observed in the SOI wafer without a surface oxide and the wafer with the bonding interface between the top-Si and BOX layer. This result indicates that the recombination via surface and interface states is effective for the decay process of the EHD luminescence. We found dependence of the EHD-PL lifetime on fabrication techniques in commercial SOI wafers with the same thickness and surface conditions, although there were negligible differences in the luminescence spectral shape among the wafers. This dependence can be explained by the difference in crystalline quality of the top-Si layer and interfacial condition of the BOX layer. We believe that the decay measurement of the EHD luminescence is a powerful technique for characterization of SOI wafers.

ACKNOWLEDGEMENTS

The authors would like to thank M. Warashina for his help in PL measurement. This work was partly supported by JSPS Research for Future Programs under the project: "Ultimate Characterization Technique of SOI Wafers for Nano-scale LSI Devices".

REFERENCES

1. M. Tajima, S. Ibuka, H. Aga and T. Abe: Appl. Phys. Lett. **70** (1997) 231.
2. M. Tajima, A. Ogura, T. Karasawa and A. Mizoguchi: Jpn. J. Appl. Phys. **37** (1998) L1199.
3. S. Ibuka, M. Tajima, M. Saito, J. Jablonski, M. Warashina and K. Nagasaka: Jpn. J. Appl. Phys. **37** (1998) 141.
4. M. Tajima and S. Ibuka: J. Appl. Phys. **84** (1998) 2224.
5. S. Ibuka and M. Tajima: Jpn. J. Appl. Phys. **39** (2000) L1124.
6. G. E. Jellison, Jr. and F. A. Modine: J. Appl. Phys. **53** (1982) 3745.
7. W. Schmid: Solid States Comm. **19** (1976) 347.
8. J. C. Cuthbert: Phys. Rev. **B1** (1970) 1552.
9. The decay of BS-2 was the fastest data of the three measured points shown in Fig. 3.